Right Living

PUBLISHING FOR THE WORLD
125 Years
THE JOHNS HOPKINS UNIVERSITY PRESS

RIGHT LIVING

*An Anglo-American
Tradition of Self-Help
Medicine and Hygiene*

EDITED BY
Charles E. Rosenberg

*Published in cooperation
with the Library Company of Philadelphia
and the College of Physicians of Philadelphia*

The Johns Hopkins University Press
BALTIMORE AND LONDON

© 2003 The Johns Hopkins University Press
All rights reserved. Published 2003
Printed in the United States of America on acid-free paper
9 8 7 6 5 4 3 2 1

The Johns Hopkins University Press
2715 North Charles Street
Baltimore, Maryland 21218-4363
www.press.jhu.edu

Library of Congress Cataloging-in-Publication Data

Rosenberg, Charles E.
    Right living : an Anglo-American tradition of self-help
medicine and hygiene / Charles E. Rosenberg.
        p. cm.
Includes bibliographical references and index.
    ISBN 08018-7189-1 (hardcover : alk. paper)
      1. Medicine, Popular—History—18th century.
2. Medicine, Popular—History—19th century.   3. Self-care,
Health—United States—History—18th century.   4. Self-care,
Health—United States—History—19th century.   I. Title.
RC81 .R797 2003
613′.0973′09033—dc21   2002007592

A catalog record for this book is available from the British
Library.

CONTENTS

Color reproductions appear between pages 116 and 117.

This book grew out of an exhibit and related conference on print culture and popular health cosponsored by two distinguished Philadelphia institutions, the Library Company of Philadelphia, founded as a subscription library in 1731, and the College of Physicians of Philadelphia, a private medical association founded in 1787. In organizing this project, these now venerable institutions were responding to late-twentieth-century phenomena: growing academic interest in the social history of medicine and in the history of books and reading. In previous generations, the history of medicine had been largely the province of physicians, few of whom were interested in medicine as everyday clinical practice or in the shifting boundaries between lay and trained practitioners. They were concerned largely with the intellectual accomplishments of the profession's elite. The history of the book, similarly, had been dominated by bibliographers and collectors, well-established communities of men and women, but groups related only tangentially to the mainstream of academic historical scholarship.

Today scores of universities and libraries boast programs or seminars in the history of the book—or print culture, as it has come to be called—while the social history of medicine has expanded well beyond a handful of historians and physicians. Younger scholars have cultivated a vigorous interest in the perspective and experience of patients and everyday practitioners—laypeople and alternative providers as well as mainstream physicians. Students of English and romance languages, of classics and literary theory, as well as cultural, demographic, and ethnohistorians have turned enthusiastically to the use of medical sources: case narratives and institutional records published and unpublished, as well as academic treatises. Not surprisingly, they have also discovered the countless books, pamphlets, and broadsides written and published for the use of laypeople in preserving their health, raising their children, and treating their ills.

But it is still a relatively neglected if richly diverse body of printed materials. No student of sexual ideas and practices, or of the complex transmission of

culture from England to the North American periphery, for example, can ignore this abundant store of texts and images. Similarly, no scholar interested
in the history of childrearing or infant mortality or of the demographic transi-
tion—that fundamental shift in patterns of birth and death associated with the
coming of the modern world—should overlook these still understudied books,
pamphlets, and broadsides. When were children weaned? How often were they
bathed? How widespread was knowledge of birth control techniques or abortion? When did ideas about cleanliness change, and how relevant were such
changed ideas to actual practice—and thus possibly to changed patterns of
morbidity and mortality? How were ideas concerning diet and exercise articulated and legitimated in terms of the preservation and restoration of health?
Obviously, the printed page is no adequate guide to behavior in all its varieties
of performance and experience, but guides to regimen and domestic practice
can provide clues to practices often so internalized—taken for granted—that
they remain invisible in more standard historical sources. Popular health texts
illuminate the practice of medicine itself and the experience of sickness, birth,
and death—all substantive aspects of human history and thus appropriate objects of scholarly investigation.

Such texts provide useful insights into worldview as well as bedside practice. Popular guides to health were—and still are—relevant to historians seeking to understand the changing uses of medicine as cultural ideology. In using
the term *cultural ideology*, I refer to a particular generation's formal ways of
thinking about the world, the body, and behavior. Which behaviors were valued? Which were stigmatized? How were the bodies of men and women physiologically constituted, and how did such speculative models legitimate behaviors and roles appropriate to the rich and to the poor, to the farmer and to the
city dweller, as well as to men and to women? How did theories of disease causation enforce such hierarchies of value, and how did they constitute templates
for proper behavior?

Print is obviously not the only source for the diffusion of such fundamental
aspects of worldview, but it is often the most important of those surviving for
the historian. And, not surprisingly, libraries and collectors have grown increasingly aware of what—in traditional rare-book terms—might be regarded
as undignified and ephemeral material. Special collections that boast of their
Vesalius and Harvey are often lacking in the less elevated works of John Gunn,
Sylvester Graham, and Samuel Thomson. Many of the subgenres of popular
health advice and admonition remain obscure, untamed by the systematic at-

tentions of the bibliographer and consigned to the underworld of ephemera. Although this situation is rapidly changing, we are still in a comparatively early stage in the exploration of this terrain.

The present book is meant to contribute to our collective understanding of these linked genres of health-oriented print. It is organized around several foci among the scores of subgenres that will ultimately attract scholarly attention; it is meant to be suggestive, not comprehensive. The first theme is the shaping—and transmission—of a much older English and continental tradition. The chapters by Steven Shapin and Mary E. Fissell illuminate several aspects of that reality: Shapin explores consensual notions of prudent regimen for the thoughtful and privileged, and Fissell the recurrent and powerful desire to understand the fabric of women's bodies and their mysterious powers of generation. Her subject, *Aristotle's Masterpiece*, was in print for a quarter of a millenium in one form or another and appealed to even the least elevated among the literate on both sides of the Atlantic. A second theme is the persistent interest in sexual relations and its consequence—childbearing and childrearing. The chapters by Kathleen Brown, Jean Silver-Isenstadt, and Ronald L. Numbers all focus on this area, as does Mary Fissell's, with its earlier chronological emphasis. A third theme is domestic practice. The chapters by Kathleen Brown and Steven Stowe address this issue, implicitly underlining as well the complexity of regional variation and the problematic relationships between text and practice. Thomas A. Horrocks and William H. Helfand address the changing issue of print as commodity and as product of specific time-bound technologies. Their subjects, almanacs and advertising posters, illuminate the variety of forms in which information about health and disease has been diffused and the marketplace motives and tactics associated with such publishing projects. Their chapters imply another truth. Media change, but medical advice and admonition soon find a place in new modes of communication. In the twentieth century, I need hardly add, film, photography, radio, television, and the Internet have all been used to disseminate an endless variety of would-be truths and suspect contentions concerning health and disease. But as I hope to have suggested, these are only the most recent modalities in which Western society has expressed the fundamental desire of men and women to manage their bodies, to maintain and restore health, and to guarantee that health to their children.

I WOULD like to acknowledge the Benjamin and Mary Siddons Measey Foundation for supporting the exhibition and symposium that inspired the present volume. I would also like to thank the Library Company of Philadelphia's Andrew W. Mellon Foundation Publication Fund for its assistance. In addition, I am in debt to James Green, William Helfand, and Thomas Horrocks for their advice in planning and to Kennie Lyman for her editorial skills.

Right Living

CHARLES E. ROSENBERG

1

Health in the Home

A Tradition of Print and Practice

When I walk into my local megabookstore, I am always struck by the diversely abundant section called "Health." It includes advice on every aspect of the human condition, from diet to incontinence, with titles devoted to an impressive variety of diseases, from depression to breast cancer, from eating disorders to lower back pain. Separate areas are organized around sexuality and addiction. And I have not even mentioned the inspirational pleas for New Age health and arguments for complementary or alternative practices. There is even a section labeled "Recovery." In a medical era dominated by intensive specialization and complex technical procedures, this proliferation of do-it-yourself health books might seem anomalous, a phenomenon of the contemporary marketplace and the alienation of today's health seekers.

It is far from that. Books and pamphlets aimed at helping men and women manage their bodies in health and disease have long been a marketable commodity. Since the beginnings of printing, readers have used the printed page to guide themselves in the preservation of physical and emotional health and in the management of their ills. The practice of popular medicine is an ancient, culturally central, and still vital reality—one that bears a complex and shifting relationship to the formal medicine of its generation. The European settlers of North America brought with them guides to regimen and midwifery as well as Bibles and prayer books. The cultural relevance of such printed artifacts has changed little over time. Many of us will, for example, have had our own lives touched by books as familiar as "Dr. Spock" or *Our Bodies, Ourselves*.[1] Today we have Web sites as well as books offering guidance in the preservation of health and management of disease, but the anxieties they address will not be assuaged. Of the making of such books there is no foreseeable end.

The American history of such self-improving works begins in early modern England, with the texts and ideas the colonists brought with them across the Atlantic—mirroring and in part embodying a centuries-old tradition of domestic practice. And it might be said to have arrived at a novel—and surprisingly modern—stage in a Civil War era of cheaper paper, printing, binding, and illustration and with the related development of increasingly national markets for books and magazines. The steam press, stereotyping, and a reliable postal service had helped shape this newly abundant and increasingly accessible universe of print. It is no accident that so many popular health books survive in shabby condition, with recipes written on the flyleaves or expanded with newspaper clippings recording cures for arthritis, consumption, or liver ailments. These books were not just read; they were used.

The Books They Brought with Them

One of the characteristic aspects of popular medical writings in the first decades after Independence was the continued dominance of British authors. The most important and widely read guides to health, regimen, and childrearing were transported like every other European cultural artifact across the Atlantic and soon reprinted in North America. The titles of some often reprinted books tell us much about their content and their reader's motives: Luigi Cornaro's *Discourses on a Sober and Temperate Life*, John Armstrong's *The Art of Preserving Health*, George Cheyne's *Essay on Health and Long Life*, S. A. Tissot's *Advice to the People in General, with regard to their Health*, and Nicholas Culpeper's *English Physician; and Complete Herbal*.

These widely owned texts displayed an extraordinary longevity. Nicholas Culpeper's works on botanic (and astrological) medicine were, for example, reprinted scores of times after their original compilation in mid–seventeenth-century England. Nineteenth-century American editions still retained the author's by then largely anachronistic astrological orientation.[2] William Buchan's *Domestic Medicine*, first printed in Edinburgh in 1769, was reprinted in scores of American editions and—though perhaps not quite so omnipresent as the

Bible—was to be found everywhere in British and American homes before the Civil War.[3] John Wesley's *Primitive Physick,* a compendium of recipes—many of them employed by the founder of Methodism himself—also circulated widely in both England and North America after its original English publication in 1747. It was reprinted as early as 1764 in Philadelphia.[4] *Aristotle's Masterpiece*—an extraordinary seventeenth-century gathering of words and woodcuts relating to birth and generation—was also widely if often surreptitiously circulated in the colonies and the new nation.[5] Continuity with older healing traditions was clear and pervasive, as exemplified concretely in the preservation and survival of these tenaciously appealing texts.

American Guides to Healing

The dominance of these English imports was gradually undermined in the first third of the nineteenth century by the growing prominence and diversity of American authors and the elaboration of reader-specific genres—books crafted for particular segments of a literate yet economically and socially diverse market; inexpensive pamphlets simply listing remedies; health manuals for seamen and plantation owners; longer, more discursive guides to regimen and therapeutics; almanacs and other advertising designed to sell particular products; manuals of midwifery and childrearing intended for women; and books aimed at helping people understand and manage their emotions, including a novel subgenre that—beginning in the 1830s and flourishing in the 1840s, 1850s, and 1860s—addressed the sexual desires and anxieties of a growing middle class. It is no accident, for example, that much of our understanding of the nineteenth-century dissemination of birth control advice, products, and rationales comes from such books and pamphlets. And the providers of popular health advice were as varied as their literary products; although the majority were written by male physicians (or individuals claiming some medical credential), some were compiled by women without such formal claims to learning, others by sectarian opponents of regular medicine, homeopaths, botanic physicians, and advocates and practitioners of hydropathy or the water cure.[6] Their motives were diverse as well. Some sought to create a physiological basis for moral reform; some sought to advertise physicians' clinical services or proprietary nostrums; some were written to earn or supplement a living by men and women who thought of themselves as writers.

Boundaries between lay and professional medical knowledge remained indistinct, certainly through the first three-quarters of the century and even later in rural areas and among the less affluent in cities. Primary care was ordinarily a lay not a medical function; a family member had to be sick indeed before consulting a physician. Every mother was expected, for example, to treat her children's coughs and colds, fevers and sprains, headaches and stomachaches. Throughout this period, moreover, preventing illness was valued as much as skill in its treatment; every aspect of life could—it was hoped—be organized so as to minimize the risk of sickness. One need not be a doctor to manage diet, the emotions, sleep, or exercise, and prevailing medical ideas about the causation and treatment of disease could in their essence be understood by any educated man or woman. All consent was informed.

Nor did physicians have the benefit of the diagnostic aids that have become routine in the twentieth century. Even the thermometer did not become an everyday part of medical practice until the last quarter of the nineteenth century. An experienced mother or grandmother could judge a fever or inspect a whitened tongue or bloodshot eye as well as any physician.[7] Therapy too was relatively straightforward (if by late-twentieth-century standards far from efficacious). Most "professional" medical therapeutics consisted of administering drugs—while the bulk of what was termed surgery consisted of managing lesions on the body's surface or setting minor breaks and fractures. Laypersons in mid–nineteenth-century America would not ordinarily have attempted to operate for removing a bladder stone, treating a cataract, or reducing a dislocated joint, but most practitioners would also have hesitated to undertake such demanding procedures. Until the last decades of the nineteenth century, voluntary surgery represented only a small part of medical practice. And many laypeople had some understanding of medical remedies as well as of diagnosis. Drugs, it should be recalled, were universally accessible to anyone who could gather herbs or pay the pharmacist or local shopkeeper (pharmacists were not the only people allowed to sell drugs). For example, until the twentieth century there were no restrictions on the purchase of opium and its derivatives or the variety of routinely used cathartics and emetics whose active ingredients were highly toxic mercury, arsenic, and antimony compounds.

The great majority of medical care was provided in the home and performed by individuals who did not think of themselves as physicians. The skills, knowledge, and responsibilities of laypersons and physicians overlapped; trained full-time physicians were in a functional sense always consultants—with the primary caregiver a family member, neighbor, or midwife. In the colo-

nial era, clergy might play the role of medical consultant in the absence of formally credentialed physicians; they were the community's only formally
trained intellectuals. And it should be recalled as well that physicians normally
treated patients in their own homes, not in the institutional settings that we
have come to think of as normal treatment sites in the twentieth century. Respectable Americans expected to receive a physician visit at home, not attend
office hours. For most of the nineteenth century, the hospital remained a marginal institution—urban and limited to the dependent and working classes. It
is not surprising that prudent families should have assumed the need for a
book explaining domestic treatment of the sick, just as they anticipated the
need for a cookbook, primer, or Bible. Families provided medical care for every
sort of ailment, from acute fevers to chronic ills such as cancer, tuberculosis,
and "dropsy." Again, publishers and booksellers were quick to provide advice
in managing the sickroom.[8]

Family members called upon a well-understood repertoire of recipes and
practices, much of it preserving and incorporating a centuries-old tradition
of—often botanic—information. Such lore was often passed on in the form of
oral tradition or carefully preserved manuscript "receipt" books—compilations of formulae and directions for everything from curing rheumatism to
tanning leather or making soap.[9] Some of it, as we have seen, came from the
texts the colonists had brought with them and those that had been reprinted in
North America. Not all of this information involved drugs; popular medicine
also included knowledge of bedside practice: how to dress cuts and wounds
and how to manage the problems of infancy and early childhood, for example.

Many titles aimed at the domestic practitioner were tools for everyday consultation, just as similar guides to legal procedure—with their model forms for
conveying land or writing wills—were used to avoid the costs and inconven-
ience of consulting a trained lawyer.[10] There was an undeniable market for
such handbooks, and at least as early as 1734 a Tidewater physician published
Every Man his own Doctor: Or the Poor Planter's Physician, the first such American guide to domestic practice. Its rambling subtitle tells us a great deal about
the assumptions and intended market of the author, John Tennent: *Prescribing
Plain and Easy Means for Persons to Cure themselves of all, or most of the Distempers, incident to this Climate, and with very little Charge, the Medicines being
chiefly of the Growth and Production of this Country.*[11]

Throughout the first century of American independence, the writing and
compiling, the publishing, and the use of books in domestic practice increased

Every Man his own *Doctor*:

OR, THE

POOR PLANTER's PHYSICIAN.

Prefcribing,

Plain and Eafy Means for Perfons to cure themfelves of all, or moft of the Diftempers, incident to this Climate, and with very little Charge, the Medicines being chiefly of the Growth and Production of this Country.

---------------------- But many Shapes
Of DEATH, and many are the Ways that lead
To his grim Cave, all difmal, yet to Senfe
More terrible at th' Entrance than within.
Some as thou faw'ft, by violent Stroke fhall dye,
By Fire, Flood, Famine, by *Intemperance* more
In Meats and Drinks, which on the Earth fhall bring,
Difeafes dire.

Paradife loft, Book XI.

The Fourth EDITION.

PHILADELPHIA:
printed and Sold by B. FRANKLIN, near the Market,
M,DCC,XXXVI.

[John Tennent], *Every Man his own Doctor* . . . (Philadelphia:
B. Franklin, 1736). Courtesy of the Historical Society of Pennsylvania.

steadily—reflecting as it did technical, economic, and cultural change. Nineteenth-century America constitutes a kind of golden age for such guides to healthful living and confident healing. Cheaper paper, postage and transportation, binding, and printing meant larger and more accessible books, while subscription sales and national distribution meant a new kind of mass market in printed consumables. Books were also more likely to be illustrated by mid–nineteenth century; both physicians and drug manufacturers used illustrated books, broadsides, and advertising to promote their goods and services. These new technical and market realities were reflected not only in the expansion of such traditional genres as guides to regimen, collections of recipes, and manuals of midwifery but also in the creation of new kinds of texts. One genre, for example, was devoted to relatively candid discussions of sex and marriage, birth control, and the "secret vice." Another concerned management of the mind and emotions. Enterprising publishers even began to offer periodicals devoted to health.[12] These books and pamphlets seem in retrospect to have been aimed at an expanding and increasingly urban middle class, at men and women concerned with defining their social identity by adopting an appropriate style of life for themselves and their children. Prebellum publishers provided an outpouring of guides to every aspect of childrearing, from infancy to adolescence.

Most striking perhaps is the proliferation of general guides to health and healing—the majority modeled in some specific or general way on Buchan's *Domestic Medicine*. Buchan itself was reprinted in a variety of competing formats, with an assortment of accompanying texts.[13] Later editions included—variously—sections on hernia, electricity, vaccination, rescue from drowning, diet for the poor, and cold bathing. Even more significant was the development in the early nineteenth century of a crop of homegrown guides to health and medical care. James Ewell's *Planter's and Mariner's Medical Companion*, which appeared in Philadelphia in 1807 and was frequently reprinted, was one of the earliest and most successful of such comprehensive family manuals.[14] By mid-century, however, John Gunn's *Domestic Medicine, or Poor Man's Friend* displaced Ewell and Buchan as America's most popular health guide. The first edition of this more assertively indigenous text appeared in Knoxville, Tennessee, in 1830. Within a decade his *Poor Man's Friend* had been reprinted more than a dozen times in a variety of small towns and cities; by 1860 it had evolved into a ponderous, ostentatiously bound subscription book.[15] Representing an energetic mix of mainstream medical teachings and frontier resourcefulness, its many reprintings indicate that Gunn succeeded in gauging the needs and assumptions of his Western and Southern readers. The salesmen who hawked such products were well aware that they faced a highly competitive marketplace; Gunn was only one among scores of authors whose books promised to bypass or supplement the physician's costly and possibly deleterious prescriptions.[16]

Sectarian Medicine

This is a period in which the medical profession's control of practice was far from absolute. In antebellum America, a number of thriving medical sects disputed the intellectual legitimacy and social authority of the regular profession: many Americans were advocates of the Thomsonian (botanic), eclectic, homeopathic, or hydropathic systems of medicine. A proliferation of specialized texts articulating their doctrines accompanied the growth of these sects; not surprisingly, all produced their own guides to domestic practice. Samuel Thomson, founder of the eponymous medical sect Thomsonianism, wrote one such as early as 1822, a *New Guide to Health*, which appeared in many subsequent editions and inspired scores of competitors over the next quarter-century.[17] All claimed to cure through the use of herbal remedies and the avoidance of regular medicine's mainstays of bleeding and "unnatural" mineral drugs. Such botanic books and pamphlets proliferated until the 1860s; although not all their authors claimed to be acolytes of Thomson's particular system, all appealed to a well-established practice of botanic healing.

This tradition had been widely disseminated in colonial America—as in England and on the Continent. In more formal and costly guise, nineteenth-century herbals maintained a continuity with an older healing and natural history tradition, while at what might be called the popular or vernacular level, cheaply printed pamphlets provided lists of remedies and their "indications." There were many competitors for this low-budget niche as the new century advanced. The most widely reprinted—and presumably used—was John Williams's so-called *Last Legacy to the People of the United States, or the Useful Family Herb Bill*, published first in 1811 and reprinted many times by 1830. The 1827 edition—a cheaply printed and sewn pamphlet of twenty-four pages—boasted that eight thousand copies had been printed in its first four months.[18] Americans with more pretensions—and dollars—could purchase full-length herbals, "embellished" with cuts of medicinally useful plants. As early as 1801, the polymath Samuel Stearns compiled and published the first domestically oriented

American manual: *The American Herbal, or Materia Medica.* In 1814 a New York competitor offered subscribers *A New and Complete American Medical Family Herbal* with plates available either hand colored or in black and white. By the 1830s, Americans could select from a variety of herbals, ranging from C. S. Rafinesque's *Medical Flora of the United States* to Wooster Beach's *American Practice of Medicine,* a generously illustrated three-volume guide to practice "on botanic principles."[19]

Homeopathic and hydropathic guides to preserving health began to appear in the 1830s and 1840s and flourished through the antebellum years. Both systems assumed an oppositional role in defining their claims vis-à-vis regular medicine—with advocates of hydropathy and homeopathy emphasizing the unnatural and physiologically debilitating aspects of regular medical practice, with its emphasis on cathartic drugs and—until midcentury—routine bleeding.[20] Logically enough, devotees of such sectarian practice were often also committed to radical health-enhancing diets and modes of life—a style of physiological reform that paralleled and in part constituted a more general reform impulse in the antebellum North. This is the period, for example, in which Sylvester Graham wrote and published widely on diet and sexual reform as well as health and medical practice; we still eat graham crackers, a vestige of Graham's comprehensive program of lifestyle management and control.[21] It is not hard to draw parallels between the secular perfectionism of such health reformers and similar efforts in more traditionally visible areas of reform such as abolition, temperance, and women's rights. Many women who attended the Seneca Falls women's rights convention in 1848 would, for example, have been familiar with Graham's works and hydropathic medicine, even if they were not themselves thoroughgoing converts.

From Marriage Bed to Cradle

Not surprisingly, the 1840s and 1850s saw the publication of a flourishing subgenre of books and pamphlets aimed at women and at relations between the sexes. Ranging from the earnest and ingenuously didactic to the sly and insinuatingly transgressive, such books promised a scientific understanding of the "physiology of generation" specifically and relations between the sexes generally.[22] Contraception was often a key theme in such books, clearly reflecting anxieties about the relationship between the proliferation of children and the ability to maintain a reassuringly respectable style of family life in America's

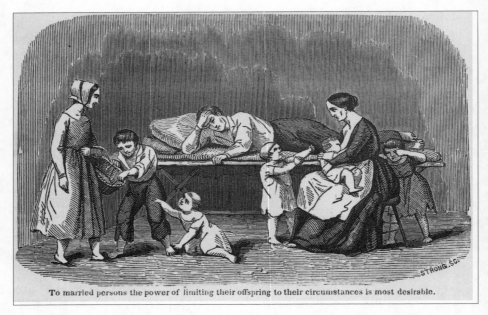

To married persons the power of limiting their offspring to their circumstances is most desirable.

Ralph Glover, *Every Mother's Book: Or the Duty of Husband and Wife Physiologically Discussed* (New York: R. Glover, 1847), frontispiece. Courtesy of the Library Company of Philadelphia.

new urban environments.[23] A peculiar subgenre of reformist health jeremiads focused on the moral and physiological dangers of masturbation. More conventional were the widely available guides to women's diseases and health and to birth, infant management, and childrearing.

Such guides had been popular in North America since at least the eighteenth century. Alexander Hamilton's *Treatise on Management of Female Complaints, and of Children in Early Infancy* was, for example, widely circulated, and his works were soon reprinted by competing American publishers.[24] Hugh Smith's *Letters to Married Women, on Nursing and Management of Children*, William Cadogan's *An Essay upon Nursing, and the Management of Children*, Michael Underwood's *Treatise on the Disease of Children*, and William Buchan's *Advice to Mothers* had similar publishing histories in this period; popular English texts, they were soon reprinted by energetic and ambitious American booksellers.[25]

By the first decade of the nineteenth century, Americans had begun to produce their own guides to child management.[26] As early as 1811, the first such text written by an American woman appeared; entitled *The Maternal Physician,* it is easily recognizable to modern eyes as a prototypical childrearing guide.[27] It has had scores of nineteenth- and twentieth-century successors; Benjamin Spock's best-selling *Common Sense Book of Baby and Child Care* was atypical largely in its extraordinary volume of sales.

In antebellum America, however, such books often served as manuals of domestic midwifery, a function now ordinarily severed from that of infant care. Today babies may be raised at home, but they are almost always born in hospitals. We no longer assume that the average household should be prepared to take responsibility for managing childbirth and the mother's aftercare. In the first half of the nineteenth century, however, midwives and general practitioners (sometimes working together) attended respectable women in their homes; only the urban, the "abandoned," and the poverty stricken would have contemplated a hospital delivery for their child. Alert authors and publishers were quick to provide printed aids.[28] Scores of books and pamphlets provided guidance for nineteenth-century American mothers in every aspect of their maternal duties—from nursing and midwifery to the management of coughs and colds.

Preserving Health

The home was important in maintaining health as well as in treating disease. In generations unaware of the germ theory and of the nature and causation of disease generally, the maintenance of health was seen in aggregate— nonspecific—terms. Every aspect of life demanded scrutiny and control. Diet, exercise, air quality, and sleep all could, over time, bring about sickness or preserve health. Books that examined life from this perspective had been written and circulated since the Renaissance; revised, reconsidered, and recycled versions of such conventional admonitions were widely read in the late eighteenth and nineteenth centuries. Formal treatises on regimen were explicitly and implicitly aimed at the "middling and affluent"; working men and women could hardly afford to alter the potentially pathogenic balance of diet, exercise, and sleep that circumscribed their lives.[29]

But anyone could become a victim of emotions—or passions, as they would have been termed at the beginning of the nineteenth century. General treatises on the preservation of health always found room for discussion of the damage uncontrolled anger, lust, or fear could inflict on the body. All medicine was psychosomatic medicine in these generations; mind and body, sickness and health were all inextricably related.[30] And by the 1820s and 1830s, a new framework for understanding human capacities and emotions had become enormously popular: phrenology, a doctrine linking the brain's localized anatomical structure with fundamental intellectual and emotional faculties and functions. In its most widely disseminated—and one might say vulgarized—version, phrenology became synonymous with the notion that psychological "readings" could be made from the contours—"bumps"—on an individual's head. Phrenological guides to health and happiness, even phrenological periodicals and almanacs, had become extraordinarily popular by the mid–nineteenth century.[31] Not all antebellum authorities on management of the mind were longtime advocates of phrenology, but all appealed to a growing American desire to understand and control mental states; consistently enough, the first English language treatises entitled "mental hygiene" were written by Americans and appeared in the midcentury United States.[32]

Continuity and Change

These books and pamphlets reflect a generally coherent vision of health and the body—even if the options for preserving health and treating disease were shaped by the realities of class and education. Of course, gender did, as we have seen, create its own readership and thus its own subgenre of advice books. Modesty, tradition, and anxiety interacted to dictate that guides to diseases of women, childbearing, and child management would flourish. But aside from a longstanding tradition of German health advice, ethnic diversity was limited.[33] Native American healing traditions were honored more by the primitivist invocations of sectarian and itinerant healers than by serious study or selective assimilation. (An assortment of hypothetical "Indian doctors" did, however, ply their marginal and often itinerant trade throughout nineteenth-century America, and Thomsonian medicine incorporated the sweat baths and herbal emetics widely associated with Native American healing practices.)

More striking than this general uniformity of content was the growing diversity of forms assumed by popular medical advice in the first half of the nineteenth century. Guidance in the prevention and cure of disease continued to fill pamphlets and books as it had in the eighteenth century, but the first half of the new century saw an increasing proliferation of novel forms through which to accomplish the traditional function of disseminating health advice. These included specialized almanacs and periodicals, newspaper columns and advertisements, and advertisements and articles in general magazines. Every technical innovation in printing and engraving was soon employed in answering an inexhaustible demand for medical information.

It is a pattern that has hardly changed. Not all of us are content to entrust our bodies to credentialed physicians and the institutions they staff and legitimate; laypeople seek assurance, understanding, and ultimately some control over their own medical prospects. The very term "managed care," which has become so familiar in recent years, embodies, on the other hand, a structured passivity; the patient is to be "managed" by experts and bureaucratic mechanisms in the patient's—and society's—presumed interest.

But as the tradition of popular health publication makes clear, men and women have always sought to manage their own care, reduce costs, and participate in the prevention and treatment of their own ills. Changes in media and markets and in the specific content of medical knowledge only underline this fundamental continuity of function. We can now access Web sites for every conceivable ill and locate support groups for sufferers and their families. And in this era of chronic disease we are, of course, deluged with every kind of advice about lifestyle as key to the avoidance and treatment of these long-term ills. As health has moved into the public sector, in addition, the desire to play an active role in health care has evolved in parallel; disease-oriented advocacy, mutual support, and lobbying groups have flourished in the late twentieth century, providing a voice for American laypersons as well as health professionals. In an era of omnipresent television and mass journalism, of films and videos, of the Internet's proliferation of words and images, of national health politics, the seemingly quaint popular medical books and pamphlets of colonial, antebellum, and Gilded Age America seem all the more alive and significant as they reflect and embody an ineradicable human desire to predict and control one's biological future.

NOTES

1. Spock's guide to child care was so widely used and influential that it soon became known by its author's name alone. For the first of many subsequent editions, see Benjamin Spock, *The Common Sense Book of Baby and Child Care* (New York: Duell, Sloan and Pearce, [1946]). Boston Women's Health Book Collective, *Our Bodies, Ourselves: A Book by and for Women* (New York: Simon and Schuster, 1973). Earlier editions were produced in a mimeographed, stapled format.

2. Culpeper's omnipresent seventeenth-century guide to English herbs—"fitted to the meanest capacity"—was reprinted as late as the 1820s; the first American edition appeared in Boston in 1708. For later editions: *The English Physician Enlarged, Containing Three Hundred and Sixty-nine Receipts for Medicines made from Herbs* (Taunton, Mass.: Samuel W. Mortimer, 1826); James Scammon, ed., *Culpepper's [sic] Family Physician: The English Physician Enlarged, Containing 300 Medicines, Made of English Herbs* (Exeter, N.H.: James Scammon, 1824).

3. Buchan was probably the most widely read of such books, appearing in almost one hundred and fifty English language editions after its original publication. Its vogue waned after the 1820s, even though it is still routinely to be found in antiquarian bookshops. Charles E. Rosenberg, "Medical Text and Social Context: Explaining William Buchan's *Domestic Medicine*," *Bulletin of the History of Medicine* 57 (1983): 22–42; C. J. Lawrence, "William Buchan: Medicine Laid Open," *Medical History* 19 (1975): 20–35. On Tissot and his *Advice to the People*, see Antoinette Emch-Deriaz, *Tissot: Physician of the Enlightenment* (New York: Peter Lang, 1992) and on George Cheyne, see Anita Guerrini, *Obesity and Depression in the Enlightenment: The Life and Times of George Cheyne* (Norman: University of Oklahoma Press, 2000).

4. A. Wesley Hill, *John Wesley among the Physicians: A Study of Eighteenth-Century English Medicine* (London: Epworth Press, 1958). After 1793, Wesley was reprinted along with the American physician Henry Wilkins's *The Family Adviser; Or, a Plain and Modern Practice of Physic; calculated for the Use of Private Families, and Accommodated to the Diseases of America . . . To which is annexed, Mr. Wesley's Primitive physic, revised* (Philadelphia: John Dickins, 1793). There were at least six subsequent American printings of this compilation before 1820, a number of these sponsored by the Methodist Church.

5. Roy Porter and Lesley Hall, "Medical Folklore in High and Low Culture: *Aristotle's Master-Piece*," in *The Facts of Life: The Creation of Sexual Knowledge in Britain, 1650–1950* (New Haven: Yale University Press, 1995), 33–64; Janet Blackman, "Popular Theories of Generation: The Evolution of Aristotle's Works: The Study of an Anachronism," in *Health Care and Popular Medicine in Nineteenth Century England*, ed. John Woodward and David Richards (New York: Holmes & Meier, 1977), 56–88; Otho T. Beall Jr., "*Aristotle's Master Piece* in America: A Landmark in the Folklore of Medicine," *William and Mary Quarterly*, 3d ser., 20 (1963): 207–22 and the chapter by Mary E. Fissell in this book.

6. There is no standard bibliography of such diverse and often ephemeral or non-mainstream publications. The University of Rochester's medical library is in the midst of publishing such an analytic bibliography. Edward G. Miner Library, *Social Medicine in the United States, 1717–1917: An Annotated and Illustrated Catalogue of the Edward Atwater Collection of American Popular Medicine and Health Reform*, compiled and annotated by Christopher Hoolihan, vol. 1, *A–L* (Rochester, N.Y.: University of Rochester Press, 2001). A second volume is in preparation.

7. And contemporary emphasis on idiosyncrasy legitimated that ability to judge deviation from the normal in a familiar face and body. Similar assumptions legitimated the logic of continuity in medical practice as well; the physician who knew all a family's members over time was—in theory—necessarily a better diagnostician and caregiver.

8. For representative examples of such guides, see, for example, Robert Wallace Johnson, *The Nurse's Guide, and Family Assistant; Containing Friendly Cautions to those who are in Health: With ample directions to nurses and others who attend the sick, women in child-bed, &c* (Philadelphia: Anthony Finley, 1818); J. S. Longshore, *The Principles and*

Practice of Nursing, or a Guide to the Inexperienced: . . . Adapted to Families, Nurses, and Young Physicians (Philadelphia: Merrihew and Thompson, 1842); Anthony T. Thomson, *The Domestic Management of the Sick-Room, Necessary, in Aid of Medical Treatment, for the Cure of Diseases. Revised, with Additions, by R.E. Griffith* (Philadelphia: Lea & Blanchard, 1845); Harriet Martineau, *Life in the Sick-Room. Essays . . . With an Introduction to the American Edition, by Eliza L. Follen,* 2d Am. ed. (Boston: William Crosby, 1845); Richard Barwell, *Guide in the Sick Room* (London: Macmillan & Co., 1864); George H. Hope, *Till the Doctor Comes, and How to Help Him. Revised, with Additions by a New York Physician* (New York: G. P. Putnam & Sons, 1871); Theodor Billroth, *The Care of the Sick at Home and in the Hospital. A Handbook for Families and for Nurses, Fourth Edition, Revised and Enlarged* (London: Sampson Low, Marston & Co.; New York: Scribners, [1894]). The terms "fever" and "dropsy" are obviously anachronistic; they do not correspond to specific disease categories as understood today. They do, however, reflect and embody categories of experience as understood in prebellum America.

9. One could also purchase such compilations, and publishers throughout the nineteenth century, especially its second and third quarters, obviously felt there was a market for books providing advice on household tasks ranging from veterinary practice to baking and dyeing—not to mention domestic care of the sick. See, for example, among scores of such books and pamphlets Prudence Smith, *Modern American Cookery . . . With a List of Family Medical Recipes, and a Valuable Miscellany* (New York: Harper & Brothers, 1835); Mrs. Rosefield and Dr. Steinburger, *"Never too Late to Learn." The Domestic Economist and Family Physician. Containing Several Hundred Valuable Receipts for Cooking Well at a Moderate expense; the Cure of Diseases; the Use of all Roots and Herbs as Medicine; Making Dyes, Coloring, Cleaning, Cementing, &c.* (New York: H. H. Randall, 1855); *Ladies' Indispensable Assistant. Being a Companion for the Sister, Mother, and Wife . . . Here are the very best directions for the Behavior and Etiquette of Ladies and Gentlemen, Ladies' Toilette Table. Directions for Managing Canary Birds. Also, Safe Directions for the Management of Children; Instructions for Ladies under Various Circumstances; A Great Variety of Valuable Recipes, forming a Complete System of Family Medicine, Thus enabling each Person to become his or her own Physician: . . .* (New York: [E. Hutchinson], 1853); E. G. Storke, ed., *The Family and Householder's Guide; Or, How to Keep House; How to Provide; How to Cook; How to Wash; How to Dye; How to Paint; How to Preserve Health; How to Cure Disease, . . . A Manual of Household Management* (Auburn, N.Y.: Auburn Publishing, ca. 1859); *Arts Revealed, and Universal Guide; Containing many rare and Invaluable Recipes and Directions for the Use of Families, from the best authorities. Embracing Directions for treating diseases—embroidery and other kinds of needlework—information as to roots and herbs—compounding medicines—how to be prepared for accidents . . .* (New York: H. Dayton, 1860); Edwin D. Freedley, ed., *Home Comforts; Or, Things worth Knowing in Every Household; . . . Respecting the Important Art of Living Well and Cheaply, Preserving Health and Prolonging Life* (Philadelphia: Claxton, Remsen & Haffelfinger, 1879). I have chosen to provide the—almost—full titles of these representative publications. They specify the content and, by implication, the audience targeted.

10. For an accessible picture of a *family's* practice in colonial Philadelphia, see Cecil K. Drinker, ed., *Not so Long Ago: A Chronicle of Medicine and Doctors in Colonial Philadelphia* (New York: Oxford University Press, 1937). For the English context, see, for example, Lucinda McCray Beier, *Sufferers and Healers: The Experience of Illness in Seventeenth-Century England* (London: Routledge & Kegan Paul, 1987); Dorothy Porter and

Roy Porter, *Patient's Progress: Doctors and Doctoring in Eighteenth-Century England* (Stanford, Calif.: Stanford University Press, 1989); and Roy Porter, ed., *Patients and Practitioners: Lay Perceptions of Medicine in Pre-Industrial Society* (Cambridge: Cambridge University Press, 1985). For some comparative perspective, see Matthew Ramsey, *Professional and Popular Medicine in France, 1170–1830: The Social World of Medical Practice* (Cambridge: Cambridge University Press, 1988); Mary Lindemann, *Health and Healing in Eighteenth-Century Germany* (Baltimore: Johns Hopkins University Press, 1996); and W. F. Bynum and Roy Porter, eds., *Medical Fringe and Medical Orthodoxy, 1750–1850* (London: Croom Helm, 1987). For a theme often prominent in tracts on popular medicine, see James C. Whorton, *Inner Hygiene: Constipation and the Pursuit of Health in Modern Society* (New York: Oxford University Press, 2000).

11. The title page describes this as the second edition; a first edition seems not to have survived (Williamsburg, Va.: William Parks, 1734). Like many such prescribing manuals, Tennent's book concluded with a handy index to diseases and another to "ingredients" that, he emphasized, were of local growth. I am "content," as he put it, "to do all my Execution with the Weapons of our own Country" (58). Colonial Williamsburg has conveniently reprinted the third edition of this rare text in an attractive facsimile.

12. The first of such periodicals was edited by Daniel Adams and was entitled *The Medical and Agricultural Register, for the Years 1806 and 1807. Containing Practical Information on Husbandry; Cautions and Directions for the Preservation of Health, Management of the Sick, &c. Designed for the Use of Families.* Printed by Manning and Loring in Boston, it ran from January 1806 to December 1807. Such serials proliferated later in the nineteenth century; sectarian movements, and in particular Thomsonian advocates of botanic medicine, were active in the prebellum years. John S. Haller Jr. has provided a valuable inventory of such periodicals: "United States Botanic Medical Journals, 1822–1860," app. H in *The People's Doctors: Samuel Thomson and the American Botanical Movement, 1790–1860* (Carbondale: Southern Illinois University Press, 2000), 271–84.

13. There was an American reprint as early as 1772. In 1775, a Philadelphia publisher issued an edition "revised and adapted to the diseases and climate of the United States of America." Rosenberg, "Medical Text," 41. In the 1790s, two competing editions "adapted to the climate and diseases of North America" went through numerous printings.

14. The title page describes Ewell as a "physician in Savannah" and is dedicated to President Thomas Jefferson. *The Planter's and Mariner's Medical Companion . . .* (Philadelphia: John Bioren, 1807).

15. Gunn's text has been conveniently reprinted: *Gunn's Domestic Medicine: A Facsimile of the First Edition* ([1830]; reprint, with an introduction by Charles E. Rosenberg, Knoxville: University of Tennessee Press, 1986). The title pages note that it was "expressly written for the benefit of families in the Western and Southern States" and contained "descriptions of the medicinal roots and herbs of the Western and Southern country, and how they are to be used in the cure of Diseases." Editions in the 1830s were published in towns like Madisonville and Pumpkintown, Tennessee, and Springfield and Xenia, Ohio. Copyright ownership changed hands, and the book's later versions appeared with big-city imprints.

16. Michael Hackenberg, "Hawking Subscription Books in 1870: A Salesman's Prospectus from Western Pennsylvania," *Papers of the Bibliographical Society of America* 78 (1984): 137–53, provides an illuminating case study of a salesman of subscriptions to one of Gunn's key rivals, Horton Howard's *Domestic Medicine: Being a Revised Edition of*

Horton Howard's Anatomy and Physiology, and Midwifery, . . . *Forming a Complete Family Medical Guide* . . . (Philadelphia: Quaker City Publishing House; San Francisco: H. H. Bancroft & Co., 1869).

17. Samuel Thomson, *New Guide to Health; Or, Botanic Family Physician* (Boston: Printed for the Author, 1822). The same year Thomson published *A Narrative of the Life and Medical Discoveries of Samuel Thomson; Containing an Account of his System of Practice, and the Manner of Curing Disease with Vegetable Medicine* . . . (Boston: The Author, 1822), which was also frequently reprinted before midcentury—sometimes along with his *New Guide.* On the Thomsonian movement, see Alex Berman and Michael A. Flannery, *America's Botanico-Medical Movements: Vox Populi* (New York: Pharmaceutical Products Press, 2001); John S. Haller Jr., *The People's Doctors;* and idem, *Medical Protestants: The Eclectics in American Medicine, 1825–1939* (Carbondale: Southern Illinois University Press, 1994). Scores of botanic guides to health appeared from American presses between the 1820s and 1860s, many published in comparatively small towns in upstate New York, New England, and Ohio. Many of their homeopathic competitors were originally English or German. See, for example, P. F. Curie, *Domestic Homeopathy. With additions and Improvements, by Gideon Humphrey* (Philadelphia: Jesper Harding, 1839); Constantine Hering, *The Homoeopathist, or Domestic Physician. 2nd American, with Additions from the 4th German Edition* (Philadelphia: Behlert and Bauersachs, 1844); John A. Tarbell, *Homoeopathy Simplified; Or, Domestic Practice made Easy* . . . *2nd edition* (Boston: Sanborn, Carter, and Bazin, 1856).

18. This twenty-four-page pamphlet was issued separately and bound with a farriery, reflecting America's still largely rural readership. Josiah Richardson, comp., *The New-England Farrier, and Family Physician; Containing, Firstly, Paul Jewett's Farriery,* . . . *Last of all, Doct. J. Williams' Family Physician* . . . (Exeter, N.H.: Josiah Richardson, 1828). The title page reassured readers that Williams was "two years with the Indians, and was assisted in that time by a young Indian educated at one of our best Medical Colleges."

19. Stearns, *The American Herbal, or Materia Medica, Wherein the Virtues of the Mineral, Vegetable, and Animal Productions of North and South America are laid open, so far as they are known; and their uses in the practice of physic and surgery exhibited* (Walpole, [N.H.]: David Carlisle for Thomas & Thomas, 1801); Samuel Henry, *A New and Complete American Medical Family Herbal, wherein is displayed the true properties and medical virtues of the plants, indigenous to the United States of America* . . . (New York: The Author, 1814); Rafinesque, *Medical Flora; Or, Manual of the Medical Botany of the United States of North America,* 2 vols. (Philadelphia: Atkinson & Alexander, 1828; Philadelphia: S. C. Atkinson, 1830); Beach, *The American Practice of Medicine: Being a Treatise on the Character, Causes, Symptoms, Morbid Appearances and Treatment of the Diseases of Women and Children,* 3 vols. (New York: Betts and Anstice, 1832). Beach's book appeared in many subsequent editions and formats; all were illustrated with botanical plates. Jacob Bigelow's landmark *American Medical Botany, Being a Collection of the Native Medicinal Plants of the United States,* 3 vols. (Boston: Cummings and Hilliard, 1817–20), with its elegantly executed and colored plates, was not aimed at the domestic practice market, although it did contain information as to the medicinal properties of the plants described. Cf. Richard J. Wolfe, *Jacob Bigelow's American Medical Botany, 1817–1821: An Examination of the Origin, Printing, Binding, and Distribution of America's First Color Plate Book* . . . (North Hills, Pa.: Bird & Bull Press; Boston: Boston Medical Library, 1979).

20. For background on these and related alternative health systems, see, for example, Martin Kaufman, *Homoeopathy in America: The Rise and Fall of a Medical Heresy* (Baltimore: Johns Hopkins University Press, 1971); Harris L. Coulter, *Divided Legacy: The Conflict between Homeopathy and the American Medical Association: Science and Ethics in American Medicine, 1800–1914,* 2d ed. (Richmond, Calif.: North Atlantic Books, 1982); Susan E. Cayleff, *Wash and Be Healed: The Water-Cure Movement and Women's Health* (Philadelphia: Temple University Press, 1987); Jane B. Donegan, *"Hydropathic Highways to Health": Women and Water Cure in Antebellum America* (Westport, Conn.: Greenwood Press, 1986).

21. See, for background, Stephen Nissenbaum, *Sex, Diet, and Debility in Jacksonian America: Sylvester Graham and Health Reform* (Westport, Conn.: Greenwood Press, 1980). For a systematic exposition of Graham's medical views, see his *Lectures on the Science of Human Life,* 2 vols. (Boston: Marsh, Capen, Lyon & Webb, 1839); for his views on sexuality and masturbation, see his *A Lecture to Young Men, on Chastity. Intended also for the Serious Consideration of Parents and Guardians* (Boston: Light & Stearns, Crocker & Brewster, 1837).

22. Frederick C. Hollick, who wrote a variety of treatises on marriage, midwifery, and the diseases of men and women, was, with the more respectable and even more prolific William A. Alcott, the most widely published—and presumably read—of authors offering advice in these vexed areas. Even today their books are easily found in secondhand bookstores and on the Internet. Thomas Low Nichols and his wife, Mary Gove Nichols, also wrote at the margin of respectability, providing their readers with a mixture of medical, lifestyle, marital, and sexual advice. In the last third of the century, New York physician and writer E. B. Foote was particularly prominent among such authors. See the chapter on Thomas Low Nichols by Jean Silver-Isenstadt and her *Shameless: The Visionary Life of Mary Gove Nichols* (Baltimore: Johns Hopkins University Press, 2002).

23. For a valuable introduction to such tracts, books, and pamphlets, see Janet Farrell Brodie, *Contraception and Abortion in 19th-Century America* (Ithaca, N.Y.: Cornell University Press, 1994). Andrea Tone's *Devices and Desires: A History of Contraceptives in America* (New York: Hill and Wang, 2001) begins with the 1870s. Norman Himes's pioneering *Medical History of Contraception* (Baltimore: Williams & Wilkins, 1936) is still useful for its documentation of early birth control literature and practices. For background, see, for example, Linda Gordon, *Woman's Body, Woman's Right: A Social History of Birth Control in America* (Harmondsworth, England: Penguin, 1977); James Reed, *From Private Vice to Public Virtue: The Birth Control Movement and American Society since 1830* (New York: Basic Books, 1978); Angus McLaren, *A History of Contraception: From Antiquity to the Present Day* (Oxford: Blackwell, 1990); idem, *Birth Control in Nineteenth-Century England* (New York: Holmes & Meier, 1978).

24. The title is itself the Edinburgh professor's own revision for "family use" of his 1781 textbook on midwifery. The enterprising printer and publisher Isaiah Thomas entitled his Worcester edition of 1793 *The Family Female Physician.*

25. The first edition of Smith's *Letters* appeared in London in 1767, the first American edition in Philadelphia in 1792. For additional details on this and other pre-1820 imprints, see Robert B. Austin, *Early American Medical Imprints: A Guide to Works Printed in the United States, 1668–1820* (Washington, D.C.: United States Public Health Service, Department of Health, Education, and Welfare, 1961).

26. See Joseph Brevit, *The Female Medical repository: To which is added, a Treatise on*

the primary diseases of infants: adapted to the use of female practitioners and intelligent mothers (Baltimore: Hunter & Robinson, 1810); Samuel K. Jennings, *The Married Lady's Companion, or Poor man's Friend; in four parts. I. An Address to the Married Lady, who is the mother of daughters. II. An address to the newly married lady. III. Some Important Hints to the Midwife. IV. An Essay on the Management and Common Diseases of Children* (Richmond, Va.: T. Nicholson, [1808]). Brevit was English, Jennings American.

27. *The Maternal Physician; A Treatise on the Nurture and Management of Infants, from the Birth until Two Years Old. Being the Result of Sixteen Years' Experience in the Nursery. Illustrated by Extracts from the Most Approved Medical Authors* (New York: Isaac Riley, 1811). Although it was published anonymously by "an American matron," the author is now known to be Mary Hunt Palmer Tyler. For a more extended discussion, see the chapter by Kathleen Brown.

28. For an attempt to create a formal midwifery curriculum, see Valentine Seaman, *The Midwives Monitor, and Mother's Mirror: being three concluding lectures of a course of instructions on midwifery* (New York: Isaac Collins, 1800). For examples of midwifery guides aimed at lay use, see George Denig, *The Domestic Instructor in Midwifery: Containing Directions for the Proper Treatment of Sexual Diseases of Women; for the Management of Pregnancy, Labor & Child-Bed; Also for the Treatment of New-Born Infants. Compiled for the advantage and use of such as have not access to a Physician* (McConnellsburg, Pa.: [The Author], 1838); Thomas Hersey, *The Midwife's Practical Directory, or Woman's Confidential Friend; Comprising Extensive Remarks on the Various Casualties, and forms of Disease, Preceding, Attending and Following the Period of Gestation, . . . Designed for the Special Use of the Botanic Friends of the United States* (Columbus, Ohio: Clapp, Gillett & Co., 1834); Horton Howard, *A Treatise on Midwifery, and the Diseases of Women and Children; Adapted for the Use of Heads of Families, and Females Particularly . . .* (Cincinnati: J. Kost, 1852); and for German-speaking readers, the often reprinted *Kurzgefasstes Weiber-Büchlein, Welches sehr Nutzlichen Unterricht fur Schwangere Weiber und Hebammen, enthalt . . .* ([Ephrata, Pa.]: [Benjamin Mayer], 1798).

29. Some full-length treatises, notably that by William Buchan, sought to bring together discursive discussions of health-enhancing regimen with specific advice on the treatment of disease; they were books to both read and use. Others focused on lifestyle in a more traditional way. See, for example, S. A. Tissot, *An Essay on Diseases incident to Literary and Sedentary Persons . . . 2nd edition* (London: J. Nourse and E. and C. Dilly, 1769); George Wallis, *The Art of Preventing Diseases and Restoring Health, founded on Rational principles, and adapted to persons of every capacity* (New York: Samuel Campbell, 1794); A. F. M. Willich, *Lectures on Diet and Regimen: Being a Systematic Inquiry into the Most Rational Means of Preserving Health and Prolonging Life . . . The 1st Boston, from the 2d London Edition* (Boston: Manning & Loring for Joseph Nancrede, 1800). For contemporary overviews and syntheses of such teachings, see Thomas Beddoes, *Hygeia: Or Essays Moral and Medical, on the Causes Affecting the Personal State of Our Middling and Affluent Classes*, 3 vols. (Bristol, England: J. Mills for R. Phillips, 1802); John Sinclair, *The Code of Health and Longevity; Or, a Concise View of the Principles Calculated for the Preservation of Health, and the Attainment of Long Life . . . The second edition*, 4 vols. (Edinburgh: Arch. Constable & Co.; London: T. Caddell, W. Davies, and J. Murray, 1807). By midcentury, efforts were being made in both England and the United States to instill healthier habits in the lower and middling orders—in schoolbooks, mechanic's institutes, and tracts and pamphlets. For an introduction to schoolbooks in this period, see

Charles E. Rosenberg, "Catechisms of Health: The Body in the Prebellum Classroom," *Bulletin of the History of Medicine* 69 (1995): 175–97.

30. George Cheyne was probably the most widely cited English language adviser in such matters. See his *The English Malady; or, a Treatise of Nervous Diseases of all Kinds, as Spleen, Vapours, Lowness of Spirits, Hypochondriacal, and Hysterical Distempers . . .* (London: G. Strahan and J. Leake, 1733) and idem, *The Natural Method of Cureing the Diseases of the Body, and the Disorders of the Mind Depending on the Body . . . The 3d edition* (London: George Strahan and J. & P. Knapton, 1742). For examples of American reprints of English texts in this psychosomatic and social critical tradition, see Thomas Trotter, *A View of the Nervous Temperament; being a Practical Inquiry into the Increasing Prevalence, Prevention, and Treatment of those Diseases Commonly called Nervous . . .* (Troy, N.Y.: Wright, Goodenow, & Stockwell, 1808); John Reid, *Essays on Hypochondriacal and other Nervous Affections . . .* (Philadelphia: M. Carey & Son, 1817). For an overview of such ideas, see Charles E. Rosenberg, "Body and Mind in Nineteenth-Century Medicine: Some Clinical Origins of the Neurosis Construct," *Bulletin of the History of Medicine* 63 (1989): 185–97.

31. The phrenological literature is enormous and reflects a widespread lay interest in managing and predicting psychological and emotional traits. See, for example, John D. Davies, *Phrenology: Fad and Science: A 19th Century American Crusade* (New Haven: Yale University Press, 1955); David de Giustino, *Conquest of Mind: Phrenology and Victorian Social Thought* (London: Croom Helm, 1975); Roger Cooter, *The Cultural Meaning of Popular Science: Phrenology and the Organization of Consent in Nineteenth-Century Britain* (Cambridge: Cambridge University Press, 1984); idem, *Phrenology in the British Isles: An Annotated, Historical Biobibliography and Index* (Metuchen, N.J.: Scarecrow Press, 1989). Mesmerism provided a parallel inspiration for popular attempts to understand and manage the mind. See, for example, Alison Winter, *Mesmerized: Powers of Mind in Victorian Britain* (Chicago: University of Chicago Press, 1998).

32. William Sweetser, *Mental Hygiene: An Examination of the Intellect and Passions, designed to illustrate their influence on health and duration of life* (New York: J. and H. G. Langley, 1843); Isaac Ray, *Mental Hygiene* (Boston: Ticknor & Fields, 1863). General books on health and regimen ordinarily included sections on the passions—we would say emotions—and their role in the etiology of disease. In the second half of the nineteenth century, moreover, books and pamphlets devoted to the "nerves and nervous" became increasingly common in the English-speaking world.

33. For a useful recent introduction to the German literature, see David L. Cowen and Renate Wilson, "The Traffic in Medical Ideas: Popular Medical Texts as German Imports and American Imprints," *Caduceus* 13 (1997): 67–80. In the colonial and early national years, the printing of German health advice was largely an eastern Pennsylvania enterprise. In the second half of the nineteenth century, mainstream English language health guides—such as those by Gunn and E. B. Foote—were translated and published in such cities as Cincinnati, St. Louis, and New York. By the end of the nineteenth century, the once abundant Pennsylvania German health imprints had largely died out.

STEVEN SHAPIN

2

How to Eat Like a Gentleman

Dietetics and Ethics in Early Modern England

> A well-behaved stomach is a great part of liberty.
> —*Montaigne (quoting Seneca)*

Consider two genres of books that were common in early modern England. One was the popular medical text. This sort of book was produced by medical practitioners and, quite often, by nonmedical men who for various reasons reckoned they had something worth saying on the subject. The general purpose of such works was to extend medical knowledge and to recommend courses of action to preserve health, cure disease, or prolong life. You could act as your own physician on many, if not all, occasions, and this genre told you how to do it, or at least reminded you of the value of what you might be presumed already to know. These books emphasized not so much diagnostics and therapeutics as dietetics—not just recommendations on what to eat and drink, but regimen and hygiene in their broadest aspects. This emphasis reflected the contemporary center of gravity in medical culture, and it also picked out a domain of action in which the maintenance of health was very much in readers' own hands, importantly taking for granted the economic ability of readers to exercise choice about their diet. Lots of these kinds of books were written from the mid-sixteenth to the eighteenth centuries, and even though we have little reliable knowledge of their circulation, ownership, and uses, we can be fairly sure that the average educated person was familiar with some of them.[1]

Another genre of popular books was composed of practical ethical tracts, including so-called courtesy books.[2] These were written by gentlemen great and small (or by "gentlemen's gentlemen"—tutors, governors, and companions) for other gentlemen who might appreciate a reminder of what the social game was all about, wanted to be recognized as gentlemen, desired that standing for their children and wished to raise them accordingly, had cultural goods to sell to gentlemanly society, or, for a variety of reasons, wished to know how gentlemen did or should behave. English courtesy books instructed readers about the authentic basis of gentility, and although they generally acknowledged that birth and wealth counted for much, they overwhelmingly stressed (alone or in various combinations) the role of virtue, education, piety, and easy good manners as the proper entitlements to gentlemanly status. Humanism and Puritanism each had their proprietary views of what the English gentleman ought to be and what was wrong with what he then was. These books explained how to live a virtuous life; how to behave in a polite, prudent, and civil manner; how to raise sons; how to pass muster; and sometimes, if necessary in a period of mask and mobility, just how to pass. There were lots of these books around too, and they were often inventoried in the emerging gentleman's "library" of the seventeenth-century English country or town house.[3] Samuel Pepys—a tailor's son, but a well-connected one, and very much on the rise, always curious about how people behaved in circles above his—was an avid consumer of such books.[4] And John Aubrey's practical thoughts on the education of gentlemen's sons recommended the reading of the better courtesy books for instilling in the young what he called "mundane prudence."[5]

Manners and medicine do not seem, on the surface of things, to have much to do with each other. Books explaining how to behave like a gentleman might be presumed very different sorts of things than books explaining how to preserve health and live long. And, indeed, from all sorts of pertinent points of view, the two genres *are* distinct: frankly medical texts do not offer rules for when to "take the wall" and when to bear your head, and a courtesy book or essay in practical deportment is unlikely to contain instructions about whether boiled or roasted meats are more suitable for an atrabilious temperament. Yet there is an overlap in substance between the two genres, and it is a telling one: that which was considered dietetically good *for* you was also accounted *morally* good. The relationship between the medical and the moral was not merely metaphorical; it was constitutive. In doing what was good for you, you were doing what was good: materially constituting yourself as a virtuous and prudent person, giving symbolic public displays of how virtuous and prudent persons behaved, encouraging such behavior in others, fulfilling the noblest aspect of your nature as a human being. The medical and the moral occupied the same terrain, figuratively in the case of cultural modes, literally in the mundane management of the body and its transactions with the world.

This cohabitation, and the consequent substantive overlap between medical culture and moral commentary, has been noted many times before, both for early modern England and more generally for premodern medicine. In the 1930s Ludwig Edelstein summarized the fundamental dictum of ancient dietetics: "He who would stay healthy must . . . know how to live rightly."[6] Owsei Temkin described the assumption of Galenic medicine that "[a] healthy life is a moral obligation . . . Health . . . becomes a responsibility and disease a matter for possible moral reflection." He beautifully depicted what it meant in antiquity to say that the philosophic life of virtue depended upon the medical care of the self, and he recounted Galen's view that "[t]he disposition of the soul is corrupted by unwholesome habits in food and drink, and in exercise, in what we see and hear, and in all the arts."[7] Temkin observed: "The coupling of right diet and virtue was an essential part of Galenic philosophy. Proper regimen balanced the temperament of the body and its parts, and with them the psychic functions. Correct and incorrect diet could determine health and disease, and because it was under human control, the choice of diet gave a moral dimension to health and sickness."[8] And on this subject Michel Foucault's account of the moral nature of ancient dietetics is little more than an expansion of insights secured by such scholars as Edelstein, Temkin, Sigerist, and a pioneering generation of students of ancient medicine.[9] More recently, Keith Thomas has briefly but perceptively written about the constitutive relationship between medicine and morality in early modern English popular medical texts, noting that "their advice coincided closely with the conventional morality of the day. Indeed, the precepts they offered were as much ethical as medical."[10] That link has also been thoroughly treated by historians dealing with the popular medical literature of later periods and in non-English settings.[11]

My aims here are modest. I want to add more depth and detail to our current understandings of the early modern English connections between dietetics and morality, but, more importantly, I want to do this by approaching that relationship from a different direction than has been customary. Supposing that you were a moral philosopher, or a historian of practical ethics, what picture of medicine would you get if you looked at historical sources central to

your discipline? What does dietetic advice look like when one encounters it not
in explicitly medical tracts but in the literature of practical ethics written by
and for gentlemen, instructing them how to live a virtuous, prudent, and effective life? What are the social-historical circumstances that mold the dietetic
counsel one finds there? What broad agreements and what contests were there
about the shape and content of this advice during the early modern period?
What does the resulting picture show about the relationship between the
layperson and the professional, between common sense and expertise? And, finally, what can this different angle of attack suggest about changing relations
between health and virtue in the early modern and in the late modern constitutions?

Nothing to Excess: Dietary Moderation and Gentlemanly Health

No English practical ethical text of which I am aware omitted passages of
medical advice, and dietetics made up by far the biggest portion of that medical advice. The dietetic counsel one finds in this literature is remarkably stable over time and setting; it reeks of prudence and robust common sense; it is
skeptical of extremes, innovations, one-size-fits-all courses of action, and
claims to external special expertise; but its very banality and cultural robust-
ness is what makes it so deeply interesting.

In 1531 Thomas Elyot's *The Governor* commended temperance in all things
and sobriety in diet: surfeit was bad for you, engendering "painful diseases and
sicknesses."[12] Thomas Gainsford's *Rich Cabinet* passed on the proverbial form:
"Temperance in diet and exercise, will make a man say; a figge, for *Gallen &
Paracelsus.*" That is to say, a temperate diet—like the apple—keeps the doctor
away.[13] When King James VI of Scotland (later James I of England) wrote to
instruct his infant son and heir, Henry, how to live like a prince, he too warned
against "using excesse of meate and drinke" and told him above all to "beware
of drunkennesse."[14] James Cleland's *The Instruction of a Young Noble-man*
(1612) followed the king's counsel closely: "[I]t is the preservation of health not
to be filled with meate; & when a man eateth more meat then his stomacke is
able to digest he becommeth sicke."[15] In the 1630s Henry Peacham's *Complete
Gentleman*—also much influenced by royal views—said the same: the gentleman was to "be moderate" in regard of his health, "which is impaired by nothing more than excess in eating and drinking (let me also add tobacco-taking).

Many dishes breed many diseases, dulleth the mind and understanding, and not only shorten but take away life."[16] Gilbert Burnet's *Thoughts on Education* even pointed to proto-eugenic reasons for shunning "all wasting intemperance, and excesse": "[S]ince the minds of children are molded into the temper of that case and body wherein they are thrust, and the healthfulness and strength of their bodies is suitable to the source and fountain from whence they spring, it clearly appears that persons wasted by drunkenness or venery must procreate unhealthful, crazy, and often mean-spirited children."[17]

Even strongly "Puritan" courtesy texts criticized contemporary dietary excess for practical medical reasons. Restoration gallants made debauchery their profession, but they would surely pay for their pleasure: "The *Table* is the *Altar* where they *sacrifice* their *Healths* to their *Appetittes;* and *Temperance* to *Luxury.*" Gentlemanly infatuation with exotic and expensive foods was proverbial: *"What's farre fetch't and deare bought is meat for Gentlemen."*[18] Jean Gailhard's *Compleat Gentleman* (1678) cautioned parents to accustom their children early to "sobriety and temperance in their diet." They should be bred to approach the table "not so much to please their palate, as to nourish their body . . . for exuberancy of food causes surfeits, which do endanger their life . . . [P]lain food is more nourishing, and less hurtful, than that which is accounted more exquisite; because the palate is pleased with it, though it be otherwise with the stomach."[19] And when John Locke, writing more as a household governor than as a philosopher, or even as an Oxford physician, composed his 1693 tract on the education of gentlemen's sons, he too advocated a "plain and simple diet." Little meat, much bread, few spices, small beer only: drunkenness, gluttony, and gormandizing to be avoided at all costs. The English ate far too much meat, and to this intemperate habit Locke imputed "a great Part" of the "Diseases in *England.*"[20] The third earl of Shaftesbury, whose early education was entrusted to Locke, later commended temperance and a moderate diet in his *Inquiry Concerning Virtue.*[21] And so said virtually all the English courtesy and practical ethical writers from the mid-sixteenth to the early eighteenth century.

English moralists' commendation of dietary moderation had a national bite to it. Continental as well as local voices of temperance reckoned that the English were tucking into far too much beef and swilling down far too much ale and that this was bad for their health. English writers of practical ethical texts, one might think, had a vested interest in criticizing contemporary dietary excess, and in judging that things were much worse than they had been in the English past. Yet these dietary jeremiads were often plausibly specific in their condemnations of *new* habits of gentlemanly intemperance, and present-day historians broadly agree with them about the facts of the matter. Anna Bryson, for example, describes the English court dinner "as a competitive exercise in conspicuous consumption"; Lawrence Stone reckons that the massive consumption of flesh—reaching mountainous proportions at the court of James I —was a possible cause of a gentlemanly plague of bladder and kidney stone; and Roy Porter and George Rousseau document eighteenth-century appreciations of a causal link between an epidemic of fashionable gout and increased English intake of meat and strong drink.[22] It was common—in England but also on the Continent—to blame new fashions in excess on the influence of foreigners. Peacham said that the English once had a national reputation for sobriety but that the imitation of Dutch and German drinking habits had ruined all that. Englishmen were now, Peacham judged, unhealthier, weaker, and even shorter than the victors of Agincourt, and this was attributed to overeating and overdrinking.[23] Seventeenth-century ethical writers identified this as an age of "luxury": not just dietary excess, but "delicacy," exoticism, variety, and complexity were often assigned to the influence of effete, debauched, and papist France and Italy.[24] And when these authors pointed to the early Stuart and Restoration court as the center of such unwholesome practices, the criticism of dietary sophistication and excess became an element in one of the major *political* conflicts of the seventeenth century: Court versus Country ideologies.

The English defenders of Good Olde Roast Beef, and lots of it, were not, of course, bereft of a response, insisting that such fare was physiologically appropriate to their damp and chilly climate; that it bred stout, hot-blooded heroes; that such straightforward and lavishly portioned victuals were suited to honest English natures; and, at a level less often surfacing in print, that dietary abundance was a mark of gentlemanly hospitality, generosity, and gusto.[25] "Gluttony was honourable," Roy Porter wrote of eighteenth-century English gentlemanly society: "[H]andsome eating was a token of success, and hospitality admired. Englishmen tucked in and took pride in their boards and bellies."[26] That great eighteenth-century Tory and gourmand, Samuel Johnson, showed just what the dietary moralists were up against when he famously announced, "He who does not mind his belly will hardly mind any thing else." Dr. Johnson at table was not a pretty sight: "[W]hile in the act of eating, the veins of his forehead swelled, and generally a strong perspiration was visible. To those whose sensations were delicate, this could not but be disgusting."[27]

Preaching and practice in ethical matters commonly diverge. So the first

Stuart king—whose court was in fact renowned for its culinary extravagance—formally commended a plain and simple diet. Nor was there anything very novel about the terms in which the English ethical writers criticized intemperance: dietary surfeit, as well as excessive variety and elaboration, had been widely identified as unhealthy and unwholesome since antiquity, and ancient medics and moralists pervasively accounted *theirs* an age of unhealthy excess.[28] As a general rule, the Golden Age of moderation tends always to lie in the past. Moreover, while English condemnations of dietary excess were shaped by local conditions, texts much read by the English that were written by foreigners and that primarily addressed foreign settings added their support. Almost no book writer in any early modern cultural context said that frequent gorging or boozing was good for the body. Erasmus's Christian Prince was to flee from "excessive drinking and eating."[29] Castiglione blandly said that it was "well known" that the ideal courtier "ought not to profess to be a great eater or drinker."[30] And the conferees of Stefano Guazzo's *Civile Conversation* "all agree[d] in blaming and condempning of them, who never cease to fill their bellies up to the throate, and whose love and lyfe consisteth in spending their time in eating and drinkeing, and in riotous and excessive gluttonie."[31] Even the cynical duc de La Rochefoucauld pointed to the prudence of dietary moderation: "[W]e would like to eat more but fear we shall be sick."[32]

When educated Englishmen read the ancients—whether in Latin, in Greek, or in English translation—they saw the continuity of the counsel of moderation. If they conceded ancient authority—and, despite the so-called moderns of the Scientific Revolution, they almost all did—then they saw in that great continuity further warrants for dietetic wisdom. Educated Englishmen could, and did, get their counsel of dietary moderation from ancient ethical and political tracts as well as from the medical writings of Hippocrates, Galen, Celsus, Pliny, and Orabasius. They could find temperance recommended in the moral writings of Aristotle, Plato, Cicero, Plutarch, and Seneca, and, indeed, many early modern ethical tracts were little more than palimpsests of such ancient sources.[33] Absolutely everywhere that bookish counsel was offered to early modern gentlemen, the Road to Wellville was signed by the Golden Mean.

The Dietetics of Virtue: Moderation and Mastery

Three sorts of Good Things happened if you observed the dietary Golden Mean. First, you preserved your health and obtained all the desiderata that depended upon health; second, you displayed your wisdom and virtue and acquired a valuable public reputation; third, you created the material conditions in your own body for enhanced virtue and wisdom. The first Good Thing is possibly still familiar to early twenty-first-century readers even if its grandmotherly common sense sets it against both the particularity and physiological detail of current voices of medical expertise and the rampant food-faddism of the culture that often pretends to owe its authority to medical expertise.

But the other two Good Things are rather less familiar to late moderns and deserve explication. Dietary moderation was a display of wisdom and prudence for several reasons. If you observed moderation, you also *showed* that you cared for your health. For a private person this was personal prudence—even when glossed by a religious idiom that made the human body God's temple—but for a public person, other matters were at stake.[34] Gentlemanly, and especially courtly, eating and drinking were overwhelmingly public acts, and they were public acts saturated with meaning. You tended to be observed as you ate and drank—at court, in the household, or in public eating and drinking places; communal eating and drinking constituted social order, displayed social order, and sent finely tuned social messages back and forth among the diners and drinkers. The "pledging of healths" followed strict rules of precedence and carried messages of desired changes in precedence. The offering of choice hunks of meat, the manner and order in which these were offered, and the conditions under which one was obliged to accept, or allowed to decline, offered morsels, were acts rich in hierarchical significance.[35] And, as we now understand through the work of Norbert Elias and his followers, the "civilizing process" of bodily control that is supposed to have done so much to configure the modern social agent was particularly visible on those occasions when gentlemen and aristocrats met to eat and drink together.[36] James I underlined for his son the political importance of resisting any temptation to eat privately: "Therefore, as Kinges use oft to eate publicklie, it is meet and honorable that ye also do so, as wel to eschew the opinion that yee love not to haunt companie, which is one of the markes of a Tyrant, as likewaies, that your delighte to eate privatlye, be not thought to be for private satisfying of your gluttonie, which ye would be ashamed should be publicklie seene."[37] "A good behaviour at Table," Gailhard wrote, "is a strong proof of a good Education." If you want to be treated like a gentleman, don't eat like a pig.[38]

King James's *Basilicon Doron* showed acute sensitivity to the obligation towards temperance that bore specially upon a prince. People inferred a king's true nature from observing his conduct at table. Accordingly, James advised his son to let his table behavior make a powerful display of self-control: "One of the publikest indifferent actiones of a King, & that manyest (especiallie strangers) wil narrowly take heed to, is, his manner of refection at his Table and his behaviour thereat." Keep your dishes simple, eat of them with restraint, and never allow yourself to succumb to gluttony or drunkenness, "whiche is a beastlie vice, namelie in a King."[39] (And if King James himself had not gone to The Globe to see Shakespeare's *Henry IV*, many of those who read *Basilicon Doron* had vividly in mind Prince Hal's youthful revels with the gluttonous John Falstaff and his cool rejection of his fat friend on assuming the throne: "Make more thy grace and less thy body hence.") These were all practical matters: on the one hand, a prince (or, indeed, any other public person) who showed that he cared little for his health also showed that he cared little for those people and public enterprises that depended upon him and his capacity for reliable, rational, and effective action. That is why the French physician Laurent Joubert said that princes had a special obligation to their health and to the dietary moderation that would secure health. First, the prince "must serve as a true model for his subjects and deputies and must be of a great perfection, more divine than human in sobriety [and] countenance." Whatever the prince does, his subjects will emulate. Second, the prince has a lot to do and may be called to decisive action at any moment. For practical reasons he cannot allow himself to be ill or incapacitated by surfeit of food or drink. Third, the prince must execute policy over extended periods of time. A long and healthy life is the material condition for effective policy, for securing succession, and for ensuring the safety and stability of the state.[40]

Such injunctions towards dietary moderation could be, and often were, conveyed in a secular medical idiom: eat moderately and you will live a long and healthy life. Or they might be cast in ripely rhetorical economic terms: Lord Burghley counseled his son towards a "plentifull" hospitality, but one kept well within "the measure of thine owne estate," for, he said, he had never encountered "any man growne poore by keeping an orderly Table"; and Josiah Dare condemned those "Epicures and Belly Gods [who] gulch down their Estates by gulps, till in the end they come to be glad of a dry Crust . . . [T]he Purses of such Prodigals may be said to be poor by their great goings on, while their Bellies may be said to be rich by their great comings in."[41] But the practical ethical literature more often spoke of temperance in frankly moral and religious language. And here it was said that the overwhelming fault of dietary excess was that it gave proof that the appetitive and bestial had gained sway over the rational, spiritual, and, therefore, uniquely human part of one's nature. That sensibility was utterly stable from the ancient Greek moralists to the seventeenth-century English ethical writers. Pagan or Christian, High or Low Church—it made little difference. Human beings were Great Amphibians: hybrid creations, partly animal and partly divine, and it was understood across a broad sweep of European culture that the self was a field of contest between rational will and appetitive desire. Accordingly, he who succumbed to excess—in food, drink, venery, or emotion—displayed a failure of rational control. The glutton or drunkard was more beast than human being. Indeed, he was worse than the beasts, "for they doe never exceed the measure prescribed by nature, but man will not be measured by the rule of his owne reason."[42] And the same hierarchy of control also followed the contours of social rank and order: the gentleman showed his entitlements through control of the appetites, while the absence of authentic gentility was displayed by the absence of restraint.[43] Dietary excess, a French ethical writer noted, "is the vice of brutish men," and Lord Burghley said that he had "never heard any commendations ascribed to a drunkard more then the well bearing of his drinke, which is a commendation fitter for a brewer's horse or a drayman, then for either a Gentleman or Servingman."[44] Authentic "generosity"—that is, in early modern usage, the virtuous essence of gentility—"teacheth men to be temperate in feeding, sober in drinking."[45]

Almost all practical ethical texts said the same sort of thing. *The Courtier* made the secular observation that temperance "brings under the sway of reason that which is perverse in our passions."[46] Elyot followed Plotinus's commendation of temperance as that which "keep[s] desire under the yoke of reason" and which permits us "to covet nothing which may be repented."[47] The Puritanical *Gentile Sinner* gave the same advice in a religious idiom: "The *Gentleman* is too much a *man* to be *without* all passion, but he is not so much a *beast* as to be governed by it." Temperance gives him "*Empire* over *himselfe*, where he gives *Law* to his *Affections*, and *limits* the extravagances of *Appetite*, and the insatiable *cravings* of *sensuality*."[48] Sir Walter Ralegh's advice to his son quantified the measure of drink along a scale leading from well-being and virtue to disease and vice: "[T]he first draught serveth for health, the second for pleasure, the third for shame, the fourth for madness."[49] Richard Lingard's avuncular counsel to a new graduate commended dietary moderation: It "discover[s] you to be your *own Master;* for he is a miserable Slave that is under the Tyranny of his Passions: and that Fountain teeming pair, *Lust* and *Rage* must especially be subdued."[50] William de Britaine's prudential guide to how to get on in Restoration society traced the causal link between political control and the display of self-control: "He who commands himself, commands the World too; and the more Authority you have over others, the more command you must have over your self."[51] And Locke's practical educational tract said "that the Principle of all Virtue and Excellency lies in a Power of denying our selves the Satisfaction of our own Desires, where Reason does not authorize them." Virtue could be acquired by practice, so dietary temperance should be practiced early.[52]

The practical ethical literature was therefore almost unanimous in its commendation of dietary moderation as a *mark* of virtue. More fundamentally, however, temperance, or deliberate moderation, was considered a virtue in itself: so said the ancients, and so said early modern ethical writers. Castiglione observed that "many other virtues are born of temperance, for when a mind is attuned to this harmony, then through the reason it easily receives true fortitude, that makes it intrepid and safe from every danger, and almost puts it above human passions."[53] Elyot wrote that the other virtues followed temperance, "as a sad and discreet matron and reverent governess," preventing excess in all other ways of being.[54] The royal herald Lodowick Bryskett noted that temperance is "the rule and measure of Vertue, upon which dependeth mans felicitie," and cited Platonic authority for the view that temperance is "the guardien or safe keeper of all human vertues."[55] Thomas Gainsford wrote that temperance is "the protectrix of all other vertues," and Richard Brathwait agreed that "no vertue can subsist without Moderation," the foundation and root of all other virtues.[56] King James gave royal warrant to the ancient hierarchy that made temperance the "Queene of all the reste" of the virtues: without self-command one could not realize any virtuous end. If virtue consisted of the Golden Mean, then temperance was literally the master virtue.[57]

The six Galenic "nonnaturals" were those forms of behavior presumed to be under volitional control whose rational management constituted the practice of traditional medical dietetics. The usual list of nonnaturals current in the early modern period included one's exposure to ambient air (the sort of place you decided to put your dwelling or spend your time), diet (in the strict sense of meat and drink), sleeping and waking, exercise and rest, retentions and evacuations (including sexual release), and the passions of the mind.[58] There is no more concrete sign of the common terrain occupied by practical ethics and practical dietetics in early modern England than the fact that several ethical texts explicitly structured their counsel through a list of the nonnaturals, while others did so implicitly or diffusely. Rationally managed moderation of the nonnaturals just was virtuous and prudent, and no special indication was, or needed to be, given that consideration had here moved from moral onto medical terrain. That is because no such cultural shift had in fact occurred.

So, for example, Peacham's practical observations on contemporary English mores listed those things upon which both health and the ability to do civic good principally depended, "which are air, eating, drinking, sleep and waking, moving and exercise, and passions of the mind: that we may live to serve God, to do our king and country service, to be a comfort to our friends and helpful to our children and others that depend on us, let us follow sobriety and temperance, and have, as Tully saith, a diligent care of our health, which we shall be sure to do if we will observe and keep that one short, but true, rule of Hippocrates: 'All things moderately and in measure.'"[59] And Locke's influential educational text started out with a series of counsels that rigorously followed the traditional list of nonnaturals.[60] Note especially that the control of the passions—or, as we would say, the emotions—counts clearly as a key item in ethical discourse. Moralists have always counseled the control of anger and of avarice, and so they did in early modern England. Yet the place of the passions in the list of Galenic nonnaturals also establishes their place at the very center of medical dietetics, regimen, and hygiene.[61] Looked at from the point of view of explicitly ethical writers, temperance in the nonnaturals seemed the soundest moral advice, while explicitly medical writers were similarly struck by the coincidence between what was good for you and what was good. In 1724 George Cheyne's *Essay of Health and Long Life* announced, "The infinitely wise *Author* of *Nature* has so contrived *Things*, that the most remarkable *Rules* of preserving *Life* and *Health* are *moral Duties* commanded us, so true it is, that *Godliness has the Promises of this Life, as well as that to come.*"[62]

Temperance is a virtue; following the dictates of temperance leads you to the Golden Mean in the observation of all other virtues; and, finally, dietary temperance creates the moral-physical conditions for virtuous thinking and acting. Virtue is circular. The circle is closed by widely shared notions—again, continuous with antiquity—about how diet influences the operation of the mind. Francis Bacon wrote, "It is certain . . . that the brain is as it were under the protection of the stomach, and therefore the things which comfort and fortify the stomach by consent assist the brain, and may be transferred to this place."[63] Other writers placed greater emphasis on the potential of dietary excess to corrupt the mind. *Sine Cerere et Baccho friget Venus* was the old adage, common in both medical and civic circles.[64] Excess in food and drink, especially of gross food and strong drink, fed both sexual desire and anger. That is why practical ethical writers could say, "Gluttonie and Drunkennes [are] the mother of al vices." The passions could not be effectively controlled, nor could the mind reason clearly, when the fires of desire and rage were stoked by dietary excess. "What operation can a minde make," Cleland asked, "when it is darkened with the thicke vapours of the braine? Who can thinke that a faire Lute filled ful with earth is able to make a sweet Harmonie? . . . [N]o more is the minde able to exercise anie good function, when the stomacke is stuffed with victuals. How ought Noble men then, whose mindes are ordained to shine before others in al vertuous and laudable actions, stop the abuse of abhominable *Epicurisme?*"[65] Dietary excess was bad for your body, but the gross blood and vapors bred by excess also "dulleth the mind and understanding."[66] By contrast, Charron said of temperance, "Neither is it serviceable to the bodie onely, but to the minde too, which thereby is kept pure, capable of wisdome and good counsell."[67] But when the mind was clouded by dietary excess, then the rational ability to control excess was compromised. The virtuous mind/body circle induced by temperance then became truly vicious. By the 1730s the fashionable physician George Cheyne, building on Hippocratic and Galenic dietetic ideas, had developed an elaborate and systematic theory of the dietary causation of melancholy—"the English malady"—a major contribution to which was excess and to which a sovereign remedy was a severe and formulaic "lowering diet."[68]

What the Mean Meant: Specifying Dietary Moderation

Like all the Aristotelian virtues, dietary moderation was poised between two vicious extremes. The practical ethical literature, however, overwhelmingly concentrated on the vices of excess, commending temperance in opposition to gluttony, drunkenness, delicacy, and overelaboration. The authors of such tracts were appearing in the person of the moralist, and the audiences they had in view were those gentlemen and aristocrats who had the resources to indulge themselves and who, in moralists' opinion, were in fact now indulging themselves on a spectacular scale. Nevertheless, there was also a minor theme in early modern ethical writing that picked out the vices attending the ascetic extreme.

Dietary asceticism was well known in early modern society; those who endorsed and embraced it spoke from some of that society's most authoritative platforms; and its cultural significance was widely understood within gentlemanly circles. Even so, dietary asceticism was very rarely advocated by ethical writers addressing themselves to civic actors, and, indeed, the practical dangers and social inconveniences of asceticism were sometimes spelled out. Asceticism was seen as strongly linked to the character of spiritual intellectuals and to patterns of disengagement and private contemplation that shaped their lives. Such asceticism and disengagement might be accorded high cultural value in civic society, but moralists generally warned gentlemen to avoid these practices. They were just not suited to the lives led by civic actors. They offended against gentlemanly obligations to generosity; they counted as a disagreeable display of self-indulgence and prissiness; they blocked the quotidian rhythms of gentlemanly social interaction; and, as I shall soon note, they might constitute both practical and social risks to the adaptability and mobility central to gentlemanly life.

If practical ethical writers would not have their readers be gluttons and drunkards, neither did they approve asceticism.[69] Robert Burton's *Anatomy of Melancholy* excoriated excess—ancient and modern—at length but more briefly noted the mischief wrought on their bodies and minds by those going to the other extreme: "too ceremonious and strict diet, being over precise, Cockney-like, curious in their observation of meats . . . just so many ounces at dinner . . . a diet-drink in the morning, cock-broth . . . [T]o sounder bodies this is too nice and most absurd." This was uncivil and unsound, but there were other dangers: monks and anchorites were well known to have driven themselves mad "through immoderate fasting."[70] Henry Peacham was one of several practical ethical writers who warned against going too far in the avoidance of dietary excess: "Neither desire I you should be so abstemious as not to remember a friend with a hearty draught, since wine was created to make the heart merry, for 'what is the life of man if it want wine?' Moderately taken, it preserveth health, comforteth and disperseth the natural heat over all the whole body, allays choleric humors, expelling the same with the sweat, etc., tempereth melancholy, and, as one saith, hath in itself a drawing virtue to procure friendship."[71]

To eat and drink like a gentleman was, then, to eat and drink both temperately and reasonably. So said virtually all the practical ethical writers of the period. The commendation of routine intemperance by any author pretending to prescribe a virtuous and prudent life is almost inconceivable. The Golden Mean was so thoroughly institutionalized in both ethical and medical canons that its denial would violate good sense and decency. The sensibility of the carnivalesque did indeed reject temperance, as did exercises of cultural subversion or inversion, but these rejections underlined just how central temperance was to early modern tradition, to orthodoxy, and to common sense.[72] Moderation was therefore a great cultural prize, and because it was such a prize, there were contests for giving its counsel specific content and for defining what moderation meant.

That content and meaning could and did vary in early modern England, yet it is noteworthy how variation, and even conflict, occurred while holding stable much or even all of the prescriptive form that counseled dietary moderation and that identified moderation as both morally good and medically good for you. The views of Francis Bacon are particularly pertinent in this connection. Bacon wrote a lot about medicine. Like several other "modern" natural philosophers of the Scientific Revolution, he considered that the medical profession was in a sorry state and that physicians' relative inability to prevent disease, to cure disease, and to extend human life were largely owing to deficiencies in physiological knowledge. "Medicine," he judged, "is a science which hath been . . . more professed than laboured, and yet more laboured than advanced."[73]

Bacon was specially unimpressed with the state of medical dietetics. Traditional advocacy of the dietary Golden Mean had become, to a degree, trite and unreflective, and it had never been informed by an adequate stock of valid empirical knowledge. Adherence to the Mean was still to count as prudence, but one must properly understand where the Mean was located and what were the nature and consequences of extremes. So Bacon—in both his essay "Of Regiment of Health" and his much longer tract on "Life and Death"—appropriated Celsus as authority for a respecification of dietary moderation. In the essay, Bacon wrote, "*Celsus* could never have spoken it as a *Physician*, had he not been a Wise Man withall; when he giveth it, for one of the great precepts of Health and Lasting; That a Man doe vary, and enterchange Contraries; But with an Inclination to the more benigne Extreme: Use Fasting and full Eating, but rather full Eating; Watching and Sleep, but rather Sleep, Sitting, and Exercise, but rather Exercise; and the like."[74] And in his philosophical work on longevity and health, Bacon similarly said, "[W]here extremes are prejudicial, the mean is the best; but where extremes are beneficial, the mean is mostly

worthless . . . [W]e should not neglect the advice of Celsus, a wise as well as a
learned physician, who advises variety and change of diet, but with an inclination to the liberal side; namely that a man should at one time accustom himself to watching, at another to sleep, but oftener to sleep; sometimes fast and
sometimes feast, but oftener feast; sometimes strenuously exert, sometimes
relax the faculties of his mind, but oftener the latter."[75] Bacon was here dis-
puting that dietary extremes *were* necessarily vicious and damaging. Often, if
not always, moving towards one extreme might be in itself more beneficial
than moving towards the other—eating a lot was better than fasting, sleeping
a lot was better than staying awake—and, when this was manifestly the case,
then the point of prudence was shifted towards—though not reaching—the
beneficial extreme. Bacon practiced what he preached: his seventeenth-
century biographer noted that Bacon followed "rather a plentiful and liberal
diet, as his stomach would bear it, than a restrained."[76] And his nineteenth-
century editors observed: "He could make nothing of a great dinner. He said,
'if he were to sup for a wager he would dine with a Lord Mayor.'"[77] Bacon not
only wrote but ate like a Lord Chancellor.

It was, Bacon recognized, fashionable to associate dietary abstinence with
longevity, and perhaps there was indeed a causal relationship of this sort: "It
seems to be approved by experience that a spare and almost Pythagorean diet,
such as is prescribed by the stricter order of monastic life, or the institutions
of hermits . . . produces longevity." But there was no absolute certainty in the
matter—later in the same tract Bacon wrote that "[f]requent fasting is bad for
longevity"—nor was it evident that longevity was the only relevant consideration. Bacon cited evidence that, among those "as live freely and in the common
way"—that is to say, civic actors—"the greatest gluttons, and those most devoted to good living, are often found the most long-lived." Nor was he aware of
any solid evidence that rigorous observance of the so-called middle diet contributed to longevity, even though it *might* be a prudent way to health. If you
really want to live according to the exact rule of moderation, then you are going
to have to do it with very great care, more care than the public actor may wish
to, or may be able to, devote to such things. More care, indeed, than it might
be worth.[78]

Celsus did indeed point out the medical benefits of a varied diet and way of
life, but Bacon went his ancient authority one better. The Mean might be defined in terms of its momentary location between dietary extremes. In this
case, the counsel of moderation would be: Drink two glasses of wine a day;

never drink ten a day; and there's no good in having any wine-free days. Alternatively, the Mean might be redefined by what we would now call a statistical distribution over time of daily behaviors. And in this case, the voice of temperance would say: Drink ten glasses of wine in a day if you wish, but don't make a habit of it. Bacon, it appears, meant to respecify dietary moderation along the latter lines. He concluded his discussion of how the body operated on aliment by offering advice at odds with dominant medical and moral counsel: "With regard to the quantity of meat and drink, it occurs to me that a little excess is sometimes good for the irrigation of the body; whence immoderate fasting and deep potations are not to be entirely forbidden."[79]

Here Bacon was articulating, and giving philosophical and gentlemanly cachet to, a dietary sensibility that evidently ran deep in lay culture, however much it was disapproved by physicians, priests, and most moralizing authors. The vomiting and purging induced by dietary excess was considered to cleanse the system, to get rid of accumulated crud and noxious substances, and to give the body a healthy catharsis. As was common in early modern gentlemanly society, Bacon himself often "took physick" for these purposes, but apparently only his personal recipe for a maceration of rhubarb in a little white wine and beer to "carry away the grosser humours of the body." Describing himself as "ever puddering in physic," Bacon acted largely as his own physician and dosed himself in moderation, with very great attention to detail.[80] Montaigne, whose essays Bacon much admired, similarly wove together medicine and manners in commending the occasional binge: "He will even plunge often into excess, if he will take my advice; otherwise the slightest dissipation will ruin him, and he will become awkward and disagreeable company."[81] Sir Thomas Browne's later compilation of commonly received errors recorded the view "[t]hat 'tis good to be drunk once a moneth, is a common flattery of sensuality, supporting it self upon physick, and the healthfull effects of inebriation."[82] Laurent Joubert deplored the popular saying "There are more old drunkards than there are old physicians," while also identifying ancient authority—Celsus again—for the advice that one should sometimes eat to surfeit.[83] And John Aubrey's life of Thomas Hobbes recorded the philosopher as saying "that he did beleeve he had been in excess in his life, a hundred times; which, considering his great age, did not amount to above once a yeare. When he did drinke, he would drinke to excesse to have the benefitt of Vomiting, which he did easily . . . but he never was, nor could not endure to be, habitually a good fellow, i.e. to drinke every day wine with company, which, though not to drunkennesse, spoiles the

Braine."[84] In the eighteenth century such lay approval of occasional excess still troubled physicians, who did not appreciate Bacon giving it further credibility. James MacKenzie's *History of Health* (1760) acknowledged that it was popularly, but falsely, attributed to Hippocrates that "getting drunk once or twice every month [w]as conducive to health."[85] In any case, such counsel was pervasive. It represents some of the sentiments that physicians were up against when they commended a rigorous observance of dietary moderation.

"To Live Physically is to Live Miserably": The Dietary Vicissitudes of the Active Life

Physicians were piqued by the respecification of moderation as occasional excess, but gentlemanly society nevertheless had excellent reasons for choosing to ignore the physicians. Bacon's essay on regimen aphoristically summed up the relevant consideration: "In *Sicknesse*, respect *Health* principally; And in *Health, Action*."[86] The early modern public actor—the gentleman, the courtier, the politician, the diplomat, the merchant, the soldier—came down firmly on the action side in the ancient debate between the *vita contemplativa* and the *vita activa*. In natural philosophy, Bacon's modernizing reforms were meant to reshape intellectual inquiry to fit the exigencies of political and economic action.[87] So, in medicine, Bacon reckoned that the legitimate test of medical practice was its ability to enhance the capacity for action. And in no case should medical counsel withdraw otherwise healthy men from the active sphere. If occasional feasting and boozing were central to the public life—and in early modern England they spectacularly were—then it could not possibly be a point of prudence or of morality to embrace medical counsel that removed public actors from those scenes in which public action occurred and in which social solidarity was made and subverted. The sort of self-indulgent discipline that was acceptable for sequestered scholars, monks, or retired gentlemen was not proper or permissible for the civic actor. La Rochefoucauld declared, "To keep well by too strict a regimen is a tedious disease in itself."[88] And when the proverbial voice similarly said that "[t]o live physically"—that is, according to the commands of physick—"was to live miserably," two things were meant: first, that it was not pleasant for your body; second, that it was a socially unacceptable way of living. The rigorous dietetics of moderation had to be tempered by other important ethical concerns, and, indeed, the meaning of moderation might even be respecified so as to align the notion of temperance with these other concerns. That is why Bacon's early biographer, giving the details of the Lord Chancellor's program of self-medication, felt obliged to insist that "he did indeed live physically, but not miserably."[89]

If you were actually ill, then, of course, you should summon your physician and accept his best advice. It was important to your part in the active life that you got better as quickly as you could. But if you were not actually ill, there was no reason to submit yourself to the severely ordered regimens prescribed by medical counsels of moderation. This principle was widely known among early modern medical and moral writers as the Rule of Celsus. In the original, the rule went like this: "A man in health, who is both vigorous and his own master, should be under no obligatory rules, and have no need, either for a medical attendant, or for a rubber and anointer. His kind of life should afford him variety."[90] The Rule of Celsus—the rule of no rule—had all the appeal of common sense; its dictates appeared both to accommodate prudential considerations and to fit with much of what counted as reliable physiological knowledge. Over a period of time—both lay and medical voices said—your body got accustomed to your usual diet and to the usual rhythms of your life. You submitted yourself at your peril to abrupt dietary change, or to the rigorous rule of dietary system. That is the sense of Seneca's maxim "A well-behaved stomach is a great part of liberty": if you have a healthy stomach, then you are not legitimately subject to any rules but those of your own normal patterns of life. Integrity is a circumstance of good digestion. Nor was it just lay public actors who fell in with the Rule of Celsus. It was such a prize that it was cited approvingly by a number of early modern physicians who elsewhere displayed their predilection for the sovereignty of expert dietary system. In these connections, physicians gave themselves considerable flexibility in specifying when patients were in fact healthy or, appearances tending to deceive, actually ill or at imminent risk of becoming so.[91]

More to the point, ethical writers dwelt extensively on why "living physically" was neither proper nor practical for the man who meant to act effectively on a public stage. King James wanted his son and heir to appreciate the link between dietary adaptability and effective rule. The consideration here was wholly political. In a face-to-face society, the showing of legitimate condescension—noblesse oblige—commonly took place at table, and this is where table manners might merge with matters of state. For physiological as well as prudential reasons you had to get used to what eating whatever was served, wherever you had to be. That is why James admonished his son to "eate in a manly, round, and honest fashion" and to get accustomed to eating "reasonable, rude and common-meates, aswel for making your body strong and durable for travel, as that ye may be the hardliner received by your meane subiects in their houses, when their cheere may suffice you." Dietary flexibility was good for you and it was good politics: "[Y]our dyet [should] bee accommodatte to your affaires, & not your affaires to your diet." When in Rome, eat as the Romans.[92] An effective prince could not allow dietary squeamishness, or his physicians' orders, to keep him away from where the action was, nor could he afford to offend potential allies and valuable followers by declining to eat, and visibly to relish, what was offered. Then as now, declining a proffered dish or drink might be taken as an act of social disengagement.[93]

Less exalted moralists fell in with the king's counsel, often citing ancient warrant for dietary flexibility. Peacham celebrated the example of the Emperor Augustus, who "was never curious in his diet, but content with ordinary and common viands. And Cato the Censor, sailing into Spain, drank of no other drink than the rowers or slaves of his own galley."[94] Locke advised that one not accustom children to regular mealtimes. A body grown used to such strict order would give trouble when public business necessitated the disruption of routine. And in no sphere of active life was such adaptability as important as in military occupations.[95] Of all gentlemanly roles, that of the soldier required a "body used to hardship," accustomed to whatever "accidents may arrive." The soldier's diet was generally unpredictable and often rough: the stomach on which, Napoleon said, an army marches had therefore to be a robust and compliant organ, tempered by the vicissitudes.[96]

So there were many reasons—the typical early modern mélange of the medical, the moral, and the prudential—why the publicly acting gentleman should not live according to expert, externally imposed, dietary rule. To be a slave to system was not civil; it was not prudent; and it was possibly even unnecessary to legitimate interests in preserving the health required to act effectively in society, and to do so until a ripe old age. The doctor's concerns were not necessarily the patient's concerns, nor should they be. Despite their pretensions, doctors didn't know it all, and they might not even know what was really pertinent to a gentleman's health and his freely chosen way of living. This was very much Montaigne's view, enormously influential in shaping Bacon's opinions on these matters, and more generally, both in French and through John Florio's translation, that of late-sixteenth- and seventeenth-century English gentlemanly society.

Like many other late Renaissance and early modern gentlemen, Montaigne preferred his own experience of his body, and its responses to regimen, to physicians' expert advice and artificial systems. The intellectual basis of being able to act as "your own physician" was adequate self-knowledge. No one could know one's body and its responses to food, drink, and patterns of living as well as oneself, but the price of such knowledge was painstaking attention. A prudent man should acquire and value such dietetic self-knowledge, and Montaigne accounted himself such a man: "I study myself more than any other subject. That is my metaphysics, that is my physics."[97] The essay "Of Experience" was composed when Montaigne was fifty-six years old—well past the age at which (as the old saying had it) a man should no longer need a physician.[98] Montaigne here deliberated upon what he had learned, how it bore upon his own proper regimen, and how he then stood in relation to external medical expertise.

"Of Experience" was an eloquent expression of Montaigne's well-known skepticism about external expertise and a vigorous defense of the moral and practical integrity that such skepticism assisted. In form, it is an essay about the general superiority of prudence to systemic pretension; in substantial content, it is in fact a dietary tract about the management of the Galenic nonnaturals. If you were a prudent person, then over the years your dietary routine had been informed by the patterns of your body's responses to food and drink, and, in turn, your body had grown accustomed to that dietary routine: "I believe nothing with more certainty than this: that I cannot be hurt by the use of things that I have been long accustomed to." Your appetite was a pretty reliable guide to what was good for you. If you liked it, it probably liked you. "I have never," Montaigne said, "received harm from any action that was really pleasant to me . . . Both in health and in sickness I have readily let myself follow my urgent appetites. I give great authority to my desires and inclinations. I do not like to cure trouble by trouble . . . My appetite in many things has of its own accord suited and adapted itself rather happily to the health of my stomach . . . It is for habit to give form to our life, just as it pleases; it is all-powerful in that; it is Circe's drink, which varies our nature as it sees fit." Even if your physicians urgently advised it, radical alteration in long-established customs could be bad for you: "Change of any sort is disturbing and hurtful." Montaigne was no fool: when his body began speaking to him in unaccustomed ways, he listened. When sharp sauces started to disagree with him, he went off them and his taste followed suit; when ill, wine lost its savor and he gave it up. But in ordinary circumstances, habit, having become a second nature, was to be respected, and

Montaigne thought that you should be careful about making abrupt changes in dietary routine unless they were absolutely necessary. Still, soldier and public actor that he once had been, Montaigne did not portray himself as a slave to habit: "I have no habit that has not varied according to circumstances"; "The best of my bodily qualities is that I am flexible"; "I was trained for freedom and adaptability." Vicissitudes of life were one thing; the sudden changes that flowed from adopting your doctor's expert systems were quite another.[99]

Montaigne saw no reason not to consult with physicians when he was genuinely ill, even if he suspected that, in general, they knew little and could do less: "The arts that promise to keep our body and our soul in health promise us much; but at the same time there are none that keep their promise less."[100] Expert physicians disagreed among themselves, and, so, if you didn't like the dietary advice that one offered, you could always pitch one doctor's favored rules against another's: "If your doctor does not think it good for you to sleep, to drink wine, or to eat such-and-such a food, don't worry: I'll find you another who will not agree with him." For that reason alone, you might as well do what you thought best, or nothing at all. The curative power of nature was, in any case, probably more effective than the art of any doctor: "We should give free passage to diseases; and I find that they do not stay so long with me, who let them go ahead; and some of those that are considered most stubborn and tenacious, I have shaken off by their own decadence, without help and without art, and against the rules of medicine. Let us give Nature a chance; she knows her business better than we do."[101]

If you put the conduct of your life under the care of physicians, Montaigne too thought they would make you miserable. Forbidding this and forbidding that, the doctors unman you and, ultimately, undo you: "If they do no other good, they do at least this, that they prepare their patients early for death, undermining little by little and cutting off their enjoyment of life."[102] It was a widely shared general ethical principle that a man who was a slave to system was less than a man. He who was ruled by others, or by a book of rules, was no free actor; he lacked the integrity central to gentlemanly identity.[103] Montaigne's essay took that general moral case and made it specific to dietetics. Change was physiologically and morally good, better and more possible for youth than age: "A young man should violate his own rules to arouse his vigor and keep it from growing moldy and lax. [There] is no way of life so stupid and feeble as that which is conducted by rules and discipline." "The most unsuitable quality for a gentleman," Montaigne declared, "is overfastidiousness and bondage to certain particular ways." Not to do, not to eat, or not to drink what was going wherever you were was "shameful": "Let such men stick to their kitchens. In anyone else it is unbecoming, but in a military man it is bad and intolerable; he . . . should get used to every change and vicissitude of life." By all means, listen to those who may have authentic medical expertise, but do not give up your freedom of action in so doing. Montaigne said that he knew of, and pitied, "several gentlemen who, by the stupidity of their doctors, have made prisoners of themselves, though still young and sound in health . . . We should conform to the best rules, but not enslave ourselves to them."[104]

One established role for philosophical knowledge in relation to medicine was its ability to guide the physician in preventing and curing disease. Skeptical of the ability of current knowledge to achieve these ends, Montaigne embraced another long-established role for philosophy: it could and should reconcile us to the inevitable circumstances of our mortal condition. To live like a man, one must learn to suffer like a man, and, finally, to die like a man: "We must meekly suffer the laws of our condition. We are born to grow old, to grow weak, to be sick, in spite of all medicine . . . We must learn to endure what we cannot avoid." And when it is time to die, one should die philosophically and well rather than live on stupidly and miserably: "Is it so great a thing to be alive?"[105]

The physicians didn't like such talk; they rarely do. Traditional medical dietetics did indeed stress the importance of self-knowledge and the dietetic practice that depended upon this self-knowledge. But physicians also insisted upon the necessity of being consulted in disease, upon the acknowledgment of their expertise in pronouncing whether patients really were well or ill, and upon their role in supervising and guiding patients' self-knowledge. So, for example, Montaigne's contemporary Joubert countered the apparently fashionable idea that there were other legitimate gentlemanly values than looking after one's health, or that submission to expert medical systems was anything but right and wise. It was, the physician wrote, very prudent to live young as if you were old, for then you really would attain vigorous old age. Or, as the proverb had it, "young old, old young." And if princes and public actors objected that they were busy and that it was better for them "to be loose and not observe any rules, schedules, or system," then Joubert settled for the best that he could get, and more than they were usually accustomed to giving. Such people should nevertheless observe "the strictest codes and rules [they] can possibly manage to apply, as much as . . . circumstances will allow."[106] The Jesuit Leonard Lessius also knew his target when he wrote of the reluctance of civic actors to take the best medical advice: "[T]he Wills and Humours of Men (we know) are stubborn and uncontroulable, and their Appetites too ungovernable to admit of any violent Restraints. Men (we see) will, at least the generality of them, eat and drink, and live according to the ordinary Course of the World, and indulge their sensual Appetites in everything to the full. Thus comes it to pass, that all their other Care and Diligence concerning these Physical maxims, or Prescripts, in the End produce little or no benefit at all." They are fools to live like this and they show their foolishness in blaming physicians for what they have brought upon themselves, "leav[ing] all entirely to Nature, and Event. To live physically they hold (according to the old Proverb) is to live miserably; and they look upon it as a very great Unhappiness for a Man to be dieted, to be denied the free Use (perhaps) of an insatiable Appetite, or Desire."[107]

In the eighteenth century even the great dietary doctor George Cheyne recognized the strength of gentlemanly moral resistance to medical rule: "The Reflection is not more common than just, That he who lives *physically* must live miserably. The Truth is, too great Nicety and Exactness about every minute Circumstance that may impair our Health, is such a Yoke and Slavery, as no Man of a generous free Spirit would submit to. 'Tis, as a *Poet* expresses it, *to die for fear of Dying.*" Yet Cheyne made his considerable living by peddling dietary advice to those well-heeled gentlemen he judged were, if not actually ill, then in imminent danger of becoming so. And therefore he tempered his appreciation of polite moral resistance by appealing to another set of ethical sensibilities also acknowledged in gentlemanly society: "But then, on the other Hand, to cut off our Days by *Intemperance, Indiscretion,* and guilty *Passions,* to live miserably for the sake of gratifying a *sweet Tooth,* or a brutal *Itch;* to die *Martyrs* to our *Luxury* and *Wantonness,* is equally beneath the Dignity of *human Nature,* and contrary to the *Homage* we owe to the *Author* of our Being. Without some Degree of *Health,* we can neither be agreeable to *ourselves,* nor useful to our *Friends.*"[108] Intemperance was at once religiously sinful, socially uncivil, and personally imprudent.

Expertise, Integrity, and Common Sense: Early and Late Modern Dietetics

If you were a poor person in the early modern period, you would act as your own physician largely because you could not afford the services of a medical expert. And if you were radically inclined, or if your material interests were at stake, you might assert the adequacy or superiority of self-treatment as a way of breaking up the medical profession's corporate power and political privileges. But if you were a free-acting gentleman in this period, neither of these considerations typically had much to do with pervasive assertions of medical self-knowledge or with skepticism towards medical expertise. Rather, you acted very substantially as your own physician because you wanted to, especially in circumstances where self-knowledge was central to medical assessments and where medical advice bore upon the fabric of everyday life and upon the identity and integrity of the self. Dietetic self-knowledge and self-care were marks of mundane prudence and moral integrity.

The management of the Galenic nonnaturals constituted the prescriptive part of early modern medical dietetics, but it also made up a substantial part of the early modern cultural practices that established personal identity and social worth. This is just a way of rephrasing the observation that medical dietetics inhabited the same cultural terrain as practical morality. And yet that cohabitation could and did give rise to conflict as well as to consensus—conflict between medical expertise and gentlemanly common sense and conflict between the goals of physicians and those of public actors. Insofar as gentlemen were not professionally qualified, they were, like the "vulgar," laypeople with respect to physicians. But their knowledge was not so easy to condemn as that of the vulgar, nor was it so easy medically to objectify them and their "conditions," nor, again, to dispute their definitions of the situations in which medical counsel might or might not have pertinence or potency.[109] Early modern gentlemen, that is to say, were laypeople of a very special sort. They had a voice, arguably more audible in literate culture than that of the physicians whom they occasionally employed. Gentlemanly prudence could not be dismissed as *mere* common sense or as meretricious "low" knowledge; gentlemanly self-knowledge was hard to gainsay; and gentlemanly goals formed a framework for evaluating physicians' advice whose legitimacy could be challenged only with great difficulty. When early modern gentlemen concurred with what physicians counseled, their assent was consequential, and when they did not, their dissent caused potentially serious problems for the credibility and the social grip of medical expertise.

What has become of that early modern dietetic common domain? One could argue that it dissolved long ago. To a considerable extent, the cultures of moral discourse and of medical expertise have gone their separate ways, though it would be very wrong to describe the divorce as absolute.[110] The adage

"You are what you eat" survives as a vestige of a largely lost dietetic culture, while the modern biochemistry of food metabolism gives the formula a renewed charge of credibility at the cost of a fundamental change in meaning. So far as the medical profession is concerned, dietetics no longer exists as a discrete subject in the curriculum. The "dietician" is widely understood to be someone arranging institutional meals for maximum nutritional content at minimum cost; the "nutrition scientist" may study metabolic pathways at many removes from the offer of practical counsel; and the heaps of dietary advice that polysaturate the common culture tend overwhelmingly to pick out the virtues or vices of specific food items in relation to specific conditions or diseases. None of these bears anything but a lexical relationship to the early modern culture of dietetics. The counsel of dietary moderation may be hard to discern in the contemporary culture of medical expertise and in what the laity seems to expect of that expertise. Quackery—defined as identifying simple explanations and remedies for complex conditions—is in the ascendant, and one could plausibly describe present-day lay attitudes to diet, disease, and health as an incoherent assemblage of discrete quackeries. The medical profession has almost wholly given up the role of counseling individuals on their way of life, save with respect to disease-specific conjunctures (for example, exercise and a low-fat diet in relation to coronary artery disease; stress reduction in connection with hypertension). Conversely, patients can rarely effectively insist that the advice of medical experts should be weighed against the range of their own life goals. The dietetic voice of moderation, insofar as it is audible at all in late modern culture, tends to come from sources other than medical experts.

Early modern gentlemanly acknowledgment that the dietetic counsel of moderation might itself have to be qualified by the demands of civility has similarly lost much of its force. If the character of the gentleman no longer exists, nevertheless the scenes of public life and social interaction that gave rise to Montaigne's and Bacon's commendation of dietary decorum still do. Yet some years ago, I gave a dinner party for eight that required—on medical and on ideological grounds—the preparation of four different menus. As to drink, two people would take no red wine, two others would take no wine at all, and one would not drink German wine. Both health and politics effectively trumped civility.

Scaled up, stripped down, and generalized, these observations about the career of dietetics are just familiar truisms about contemporary culture. They bear a family resemblance to academic cultural-theoretical truisms about modernity that plausibly talk about cultural specialization and differentiation, about the dominance of expertise, about secularization, about the hollowing out and debasement of a once common culture, and about the decline of civility. But, like many truisms about the nature of the modern condition, they tend to mistake the part for the whole, an aspect of how our society is changing for an adequate description of what it is. Indeed, dietetics—interestingly poised at the intersection of self and nonself, of the scientifically descriptive and the morally prescriptive—offers a perspicuous site for reconsidering how one might go about describing the modern condition.

It is worth just pointing out the conditions in which, and the extent to which, early modern dietetic culture remains intact as we enter the twenty-first century. Much of our personal identity and social worth is still asserted, established, and contested through the personal management of the Galenic nonnaturals and through others' observations of how we manage them. To that extent, it might be said that medicine and morality continue their cohabitation. Gluttony, drunkenness, sloth, promiscuity, and repeated outbursts of road rage are still likely to be black marks in assessments of individual character, and, although the counsel severely lacks trendiness, adherence to the Golden Mean probably remains a prudent course of action, even now, for anyone wishing to win friends, to influence people, and, for that matter, to decide what and how to eat and drink. Even after hard reflection, and with due deference to the physicians, the psychotherapists, the social workers, and the personal trainers, it is difficult to know what better advice one could possibly give. Of course, we no longer *know* that these practices have a coherent medical identity, and therefore, for us, they do *not* have a medical identity. Those of us who have passed through physiology courses in school or university are unlikely to credit the humoral framework that linked the emotions to diet and that offered a conceptual framework explaining why moderation is good for you.

So dietetic culture, one might say, does survive in late modernity, but as a dispersed set of fragments, ripped free from its original moorings in a medical understanding of the self and without any substantial anchorage in contemporary medical expertise. Like many other allegedly "premodern" modes, the dietetic culture that once bound medicine and morality is possibly better described as submerged in the layered streams of late modern life rather than as dissolved in a unitary, and universally credible, medical expertise. We late moderns resemble Montaigne in many things, not least in our ability to pick and choose which, if any, among the many dissenting voices of medical expertise we will listen to and act upon. Expertise without credibility is nothing at all.

And, while the counsel of dietary moderation is indeed hard to hear in modern culture, it is far from impossible to hear it, and, on hearing it, to be reminded of its sheer common sense and unassailable authority. As I finish this chapter, the current issue of the *New Yorker* brings me Julia Child's irresistible, and only slightly modernized, expression of the ancient Rule of Celsus: "A low-fat diet . . . What does that mean? I think it's unhealthy to eliminate things from your diet. Who knows what they have in them that you might need? Of course, I'm addressing myself to normal, healthy people . . . What I'm trying to do is encourage people to embrace moderation—small helpings, no seconds, no snacking, a little bit of everything, and above all, have a good time."[111]

NOTES

1. Some examples from the Tudor to the early Hanoverian period include Thomas Elyot, *The Castel of Helthe* (1539); John Goeurot, *The Regiment of Life* (1546); Andrewe Boorde, *Compendyous Regyment or a Dyetary of Healthe* (1547); Luigi Cornaro, *The Temperate Life* (1558–66; common in English translations of the Italian original); Thomas Cogan, *The Haven of Health* (1589); Leonard Lessius, *Hygiasticon* (ca. 1600; again in translation); William Vaughan, *Naturall and Artificial Directions for Health* (1602); Sir John Harington's verse translation of the *Regimen Sanitatis Salernitanum* (as *The Englishmans Doctor*) (1607), and several other prior translations of the same; the anonymous *The Skilful Physician* (1656); Nicholas Culpeper, *A Physicall Directory* (1649); John Archer, *Every Man His Own Doctor* (1671); Thomas Tryon, *The Way of Health, Long Life and Happiness* (1683); George Cheyne, *An Essay of Health and Long Life* (1724). Studies of the early modern English popular medical literature notably include Paul Slack, "Mirrors of Health and Treasures of Poor Men: The Use of the Vernacular Medical Literature of Tudor England," in *Health, Medicine and Mortality in the Sixteenth Century*, ed. Charles Webster (Cambridge: Cambridge University Press, 1979), 237–73; Andrew Wear, "The Popularization of Medicine in Early Modern England," in *The Popularization of Medicine, 1650–1850*, ed. Roy Porter (London: Routledge, 1992), 17–41; John Henry, "Doctors and Healers: Popular Culture and the Medical Profession," in *Science, Culture, and Popular Belief in Renaissance Europe*, ed. Stephen Pumfrey, Paolo L. Rossi, and Maurice Slawinski (Manchester: Manchester University Press, 1991), 191–221, esp. 198–201; Henry E. Sigerist, "The *Regimen Sanitatis Salernitanum* and Some of Its Commentators," in *Landmarks in the History of Hygiene* (London: Oxford University Press, 1956), 20–35; Charles Webster, *The Great Instauration: Science, Medicine, and Reform, 1626–1660* (London: Duckworth, 1975), chap. 4 (for the ideological and political context of popular medical writing); and, for a fine recent survey of British and Continental dietetics, Heikki Mikkeli, *Hygiene in the Early Modern Medical Tradition*, Annals of the Finnish Academy of Sciences and Letters, Humaniora, no. 305 (Saarijärvi, Finland: Academia Scientiarum Fennica, 1999). Unavailable to me at the time of writing this chapter was Ken Albala,

Eating Right in the Renaissance (Berkeley: University of California Press, 2002), a study of European dietetic books from the mid-fifteenth to the mid-seventeenth centuries.

2. For studies of the courtesy and related genres, see Anna Bryson, *From Courtesy to Civility: Changing Codes of Conduct in Early Modern England* (Oxford: Oxford University Press, 1998); Frank Whigham, *Ambition and Privilege: The Social Tropes of Elizabethan Courtesy Literature* (Berkeley: University of California Press, 1984); John E. Mason, *Gentlefolk in the Making: Studies in the History of Courtesy Literature and Related Topics from 1531 to 1774* (1935; reprint, New York: Octagon Books, 1971); George C. Brauer Jr., *The Education of a Gentleman: Theories of Gentlemanly Education in England, 1660–1775* (New York: Bookman Associates, 1959); W. L. Ustick, "*Advice to a Son:* A Type of Seventeenth Century Conduct Book," *Studies in Philology* 29 (1932): 409–41; idem, "Changing Ideals of Aristocratic Character and Conduct in Seventeenth-Century England," *Modern Philology* 30 (1933): 147–66; Fenela Ann Childs, "Prescriptions for Manners in English Courtesy Literature, 1690–1760, and Their Social Implications" (D. Phil. diss., Oxford University, 1984); Gertrude Elizabeth Noyes, *Bibliography of Courtesy and Conduct Books in Seventeenth-Century England* (New Haven: Tuttle, Morehouse & Taylor, 1937); and Peter Burke, *The Fortunes of the "Courtier": The European Reception of Castiglione's "Cortegiano"* (Oxford: Polity Press, 1995).

3. For example, J. T. Cliffe, *The World of the Country House in Seventeenth-Century England* (New Haven: Yale University Press, 1999), 166–67.

4. Of Francis Osborne's *Advice to His Son*, Pepys said, "I shall not never admire it enough for sense and language." *The Diary of Samuel Pepys*, ed. Robert C. Latham and William Matthews, 11 vols. (London: HarperCollins, 1995), 4:96 (5 April 1663); see also 2:199 (19 October 1661), 5:10 (9 January 1663/64), and 5:27 (27 January 1663/64). The decidedly ungenteel experimentalist Robert Hooke possessed four or five courtesy books, though they did not notably affect either his deportment or his diet: Leona Rostenberg, *The Library of Robert Hooke: The Scientific Book Trade of Restoration England* (Santa Monica, Calif.: Modoc Press, 1989), 198, 205–6; Steven Shapin, "Who Was Robert Hooke?" in *Robert Hooke: New Studies*, ed. Michael Hunter and Simon Schaffer (Woodbridge, England: Boydell Press, 1989), 253–85, quotation on 276.

5. Aubrey's selection included works by Peacham, Osborne, Ralegh, de Courtin, Shaftesbury, Montaigne, and Bacon: *Aubrey on Education: A Hitherto Unpublished Manuscript by the Author of "Brief Lives,"* ed. J. E. Stephens (London: Routledge & Kegan Paul, 1972; comp. from 1669 to ca. 1694), 131–32.

6. Ludwig Edelstein, "The Dietetics of Antiquity," in idem, *Ancient Medicine: Selected Papers of Ludwig Edelstein*, ed. Owsei Temkin and C. Lilian Temkin (Baltimore: Johns Hopkins Press, 1967; article originally published 1931), 303–16, quotation on 303.

7. Galen, *De sanitate tuenda*, quoted in Owsei Temkin, *Galenism: Rise and Decline of a Medical Philosophy* (Ithaca, N.Y.: Cornell University Press, 1973), 39–40.

8. Owsei Temkin, *Hippocrates in a World of Pagans and Christians* (Baltimore: Johns Hopkins University Press, 1991), 15, 45–47. See also Sigerist, "Galen's *Hygiene*," in *Landmarks in the History of Hygiene*, 1–19, esp. 12.

9. Michel Foucault, *The History of Sexuality*, trans. Robert Hurley, 3 vols. (New York: Vintage Books, 1988–90), 2:97–139; see also 3:140–41. (In these connections, Foucault did not cite Edelstein et al., nor, in the approved French academic manner, did he acknowledge the work of any other living or recently deceased scholar. But it is hard to

imagine that, as his references imply, Foucault relied solely on ancient primary sources for his knowledge of dietetics.) For present-day classicists' and moral philosophers' reflections on ethics and body management in antiquity, see also Martha Nussbaum, *The Therapy of Desire: Theory and Practice in Hellenistic Ethics* (Princeton, N.J.: Princeton University Press, 1994); John Cottingham, *Philosophy and the Good Life: Reason and the Passions in Greek, Cartesian, and Psychoanalytic Ethics* (Cambridge: Cambridge University Press, 1998); Alexander Nehamus, *The Art of Living: Socratic Reflections from Plato to Foucault* (Berkeley: University of California Press, 1998); and especially James Davidson, *Courtesans and Fishcakes: The Consuming Passions of Classical Athens* (London: HarperCollins, 1997).

10. Keith Thomas, "Health and Morality in Early Modern England," in *Morality and Health*, ed. Allan M. Brandt and Paul Rozin (New York: Routledge, 1997), 15–34, quotation on 20. See also idem, *Man and the Natural World: Changing Attitudes in England, 1500–1800* (London: Allen Lane, 1983), 289–300 (esp. for vegetarian dietetics and ethics); and Slack, "Mirrors of Health." Unavailable to me at the time of writing this chapter was Pelling's fine essay "Food, Status, and Knowledge: Attitudes to Diet in Early Modern England," in *The Common Lot: Sickness, Medical Occupations, and the Urban Poor in Early Modern England*, ed. Margaret Pelling (London: Longman, 1998), 38–62.

11. For example, William Coleman, "Health and Hygiene in the *Encyclopédie*: A Medical Doctrine for the Bourgeoisie," *Journal of the History of Medicine* 29 (1974): 399–421; idem, "The People's Health: Medical Themes in 18th-Century French Popular Literature," *Bulletin of the History of Medicine* 51 (1977): 55–74; Christopher J. Lawrence, "William Buchan: Medicine Laid Open," *Medical History* 19 (1975): 20–36; Charles E. Rosenberg, "Medical Text and Social Context: Explaining William Buchan's *Domestic Medicine*," *Bulletin of the History of Medicine* 57 (1983): 22–42; idem, "The Therapeutic Revolution: Medicine, Meaning, and Social Change in Nineteenth-Century America," *Perspectives in Biology and Medicine* 20 (1977): 485–506; idem, "Florence Nightingale on Contagion: The Hospital as Moral Universe," in *Healing and History: Essays for George Rosen* (New York: Science History Publications, 1982), 116–36; Ginnie Smith, "Prescribing the Rules of Health: Self-Help and Advice in the Late Eighteenth Century," in *Patients and Practitioners: Lay Perceptions of Medicine in Pre-Industrial Society*, ed. Roy Porter (Cambridge: Cambridge University Press, 1985), 249–82.

12. Thomas Elyot, *The Book Named The Governor*, ed. S. E. Lehmberg (1531; reprint, London: Dent, 1962), 214.

13. [Thomas Gainsford], *The Rich Cabinet furnished with varieties of Excellent Discriptions, Exquisite Characters, Witty Discourses, and Delightfull Histories, Devine and Morrall . . . Whereunto is Annexed the Epitome of Good Manners, extracted from Mr. John de la Casa, . . .* (1616; reprint, Amsterdam: Da Capo Press for Theatrum Orbis Terrarum, 1972), 143v.

14. James VI, King of Scotland, later James I of England, *Basilicon Doron*, facsimile ed. (1599; reprint, Menston, England: Scolar Press, 1969), 126.

15. James Cleland, *The Instruction of a Young Noble-man* (Oxford, 1612), 211. Large sections of Cleland's tract are just (unacknowledged) quotations or close paraphrases of the king's *Basilicon Doron*.

16. Henry Peacham, *The Complete Gentleman*, ed. Virgil B. Heltzel (1622, 1634; reprint, Ithaca, N.Y.: Cornell University Press, for the Folger Shakespeare Library, 1962), 151–52.

17. Gilbert Burnet, *Thoughts on Education*, ed. John Clarke (Aberdeen: Aberdeen University Press, 1914; posthumously published 1761; comp. ca. 1668), 13. Burnet concurred with common early modern hereditary sentiment in urging that care be taken in selecting children's wet nurses for their dietary temperance and good moral character (14–15).

18. Clement Ellis, *The Gentile Sinner, or England's Brave Gentleman Character'd in a Letter to a Friend: Both As He Is, and As He Should Be*, 4th ed. (1660; reprint, Oxford: 1668), 193.

19. Jean Gailhard, *The Compleat Gentleman: Or Directions for the Education of Youth*, 2 pts., separately paginated (London, 1678), pt. 1, 88.

20. John Locke, *Some Thoughts Concerning Education*, ed. R. H. Quick (1693; reprint, Cambridge: Cambridge University Press, 1899), 9–12.

21. Anthony Ashley Cooper, 3d earl of Shaftesbury, *An Inquiry Concerning Virtue, in Two Discourses . . .* (London, 1699), 162–73. See also Henry Richard Fox Bourne, *The Life of John Locke*, 2 vols. (1876; reprint, Darmstadt, Germany: Scientia Verlag Aalen, 1969), 2:255–64.

22. Bryson, *From Courtesy to Civility*, 121; Lawrence Stone, *The Crisis of the Aristocracy, 1558–1641* (Oxford: Clarendon Press, 1965), 555–62; Roy Porter and G. S. Rousseau, *Gout: The Patrician Malady* (New Haven: Yale University Press, 1998), esp. 48–59. See also Roy Porter, *English Society in the Eighteenth Century* (Harmondsworth, England: Penguin, 1982), 3–35, 233–35; Andrew B. Appleby, "Diet in Sixteenth-Century England: Sources, Problems, Possibilities," in *Health, Medicine, and Mortality in the Sixteenth Century*, ed. Webster, 97–116; and Peter Earle, *The Making of the English Middle Class: Business, Society, and Family Life in London, 1660–1730* (Berkeley: University of California Press, 1989), 272–81. Earle estimates that even those Londoners in "the middle station" of society consumed meat four or five days a week, and the modern Californian runs the risk of furred arteries by just reading the carnivorous bill of fare for gentlemanly feasts.

23. Peacham, *Complete Gentleman*, 153, 235–36, 238–39.

24. Richard Lingard, *A Letter of Advice to a Young Gentleman Leaving the University Concerning His Behaviour and Conversation in the World*, ed. Frank C. Erb (1670; reprint, New York: McAuliffe & Booth, 1907), 41. See also Thomas, "Health and Morality," 21, and Burke, *Fortunes of the "Courtier,"* 113–15.

25. In the 1540s Sir Thomas Elyot celebrated the nutritional virtues of English roast beef, while recognizing the possible dangers of excess: "Biefe of Englande to Englysshemen, whiche are in helth, bringeth stronge nouryshynge, but it maketh grosse bloude, and ingendreth melancoly." *The Castel of Helthe* (London, 1541), 16r.

26. Porter, *English Society in the Eighteenth Century*, 34–35.

27. James Boswell, "The Life of Samuel Johnson," in *The Portable Johnson & Boswell*, ed. Louis Kronenberger (New York: Viking, 1947), 122–23.

28. On this, see the learned and entertaining Davidson, *Courtesans and Fishcakes*.

29. Desiderius Erasmus, *The Education of a Christian Prince*, trans. Lester K. Born (1516; reprint, New York: Octagon, 1965), 209.

30. Baldesar Castiglione, *The Book of the Courtier*, trans. Charles S. Singleton (1528; reprint, Garden City, N.Y.: Doubleday Anchor, 1959), 135.

31. Stefano Guazzo, *The Civile Conversation of M. Steeven Guazzo*, trans. George Pettie and Barth[olomew] Young, ed. Sir Edward Sullivan, 2 vols. (1574 [Italian], 1581, 1586 [English]; reprint, London: Constable and Co., 1925), 2:137.

32. François, duc de La Rochefoucauld, *The Maxims of La Rochefoucauld*, trans. Louis Kronenberger (1665; reprint, New York: Random House, 1959), 142.

33. See, among very many such ancient arguments for dietary moderation influential in early modern England, Plutarch, "Rules for the Preservation of Health," in *Plutarch's Lives and Miscellanies*, ed. A. H. Clough and William W. Goodwin, 5 vols. (New York: Colonial Company, 1905), 1:251–79; and "Plutarch's Symposiacs," in ibid., 3:197–460, esp. 290–95, 339, 394–98. Of course, the average educated Englishman—even equipped with decent school Latin—was more likely to encounter both the ethical and the medical knowledge of antiquity via early modern English summaries and compendia.

34. On the body as divine temple in English Protestant thought, see, for instance, Keith Thomas, "Cleanliness and Godliness in Early Modern England," in *Religion, Culture, and Society in Early Modern Britain: Essays in Honour of Patrick Collinson*, ed. Anthony Fletcher and Peter Roberts (Cambridge: Cambridge University Press, 1994), 56–83, esp. 62–63; idem, "Health and Morality," 16–18.

35. Giovanni Della Casa, *Galateo*, trans. Konrad Eisenbichler and Kenneth R. Bartlett (1558, widely available in English translation from 1576; reprint, Toronto: Centre for Reformation and Renaissance Studies, 1986), 9–10, 57–59; Antoine de Courtin, *The Rules of Civility; or, Certain Ways of Deportment observed amongst all Persons of Quality upon Several Occasions, newly revised and much enlarged* (1671; reprint, London: 1685), 122–45; Peacham, *Complete Gentleman*, 153; Ellis, *Gentile Sinner*, 189; Gailhard, *Compleat Gentleman*, pt. 1, 90; Josiah Dare, *Counsellor Manners: His Last Legacy to His Son* (1672; reprint, New York: Coward-McCann, 1929), 16–18; John Evelyn, *A Character of England*, in *Harleian Miscellany*, ed. T. Park (London, 1808–13), 10:189–98; Cleland, *Instruction of a Young Noble-man*, 211–12; Bryson, *From Courtesy to Civility*, 83, 93, 121; idem, "The Rhetoric of Status: Gesture, Demeanour, and the Image of the Gentleman in Sixteenth- and Seventeenth-Century England," in *Renaissance Bodies: The Human Figure in English Culture c. 1540–1660*, ed. Lucy Gent and Nigel Llewellyn (London: Reaktion Books, 1990), 136–53, on 145, 150–51.

36. Norbert Elias, *The Civilizing Process*, trans. Edmund Jephcott, 2 vols. [vol. 1 = *The History of Manners*; vol. 2 = *The Court Society*] (Oxford: Basil Blackwell, 1978, 1983), esp. vol. 1, chap. 2, pt. 4 ("On Behavior at Table"). See also Stephen Mennell, *All Manners of Food: Eating and Taste in England and France from the Middle Ages to the Present* (Oxford: Basil Blackwell, 1985); idem, "On the Civilizing of Appetite," in *The Body: Social Process and Cultural Theory*, ed. Mike Featherstone, Mike Hepworth, and Bryan S. Turner (London: Sage, 1991), 126–56; Janet Whatley, "Food and the Limits of Civility," *Sixteenth Century Journal* 15 (1984): 387–400; cf. the historical criticisms of Elias in Bryson, *From Courtesy to Civility*, 105. Cultural anthropologists, of course, have made a meal of food, and food-giving, symbolism; see, among very many examples, Mary Douglas and Jonathan L. Gross, "Food and Culture: Measuring the Intricacy of Rule Systems," *Social Science Information* 20 (1981): 1–35; Jack R. Goody, *Cooking, Cuisine, and Class* (Cambridge: Cambridge University Press, 1982); Claude Lévi-Strauss, *The Raw and the Cooked*, trans. John and Doreen Weightman (1964; reprint, New York: Harper & Row, 1969); idem, *From Honey to Ashes*, trans. John and Doreen Weightman (1966; reprint, New York: Harper & Row, 1973); idem, *The Origin of Table Manners*, trans. John and Doreen Weightman (1968; reprint, New York: Harper & Row, 1978). So, too, have cultural historians, for instance, Caroline Walker Bynum, *Holy Feast and Holy Fast: The Religious Significance of Food to Medieval Women* (Berkeley: University of California Press,

1987); Peter Brown, *The Body and Society: Men, Women and Sexual Renunciation in Early Christianity* (London: Faber and Faber, 1989); Piero Camporesi, *Bread of Dreams: Food and Fantasy in Early Modern Europe*, trans. David Gentilcore (Cambridge: Polity Press, 1989).

37. King James, *Basilicon Doron*, 124.

38. Gailhard, *Compleat Gentleman*, pt. 2, 67–68. See also Cleland, *Instruction of a Young Noble-man*, 210; Della Casa, *Galateo*, 57; de Courtin, *Rules of Civility*, 131–33; Thomas, "Health and Morality," 27.

39. King James, *Basilicon Doron*, 124, 126. Here, as elsewhere, Cleland closely followed royal advice: *Instruction of a Young Noble-man*, 207–13. See also Erasmus, *Education of a Christian Prince*, 209, and Sir John Harington's popular verse translation of the Salernitan canon: "A King that cannot rule him in his dyet, / Will hardly rule his Realme in peace and quiet." *The School of Salernum: Regimen Sanitatis Salerni* (1607; reprint, Salerno: Ente Provinciale per Il Turismo, 1957), 50. (Harington was a favorite of Henry, Prince of Wales, and Cleland dedicated his *Instruction* to Harington.)

40. Laurent Joubert, *The Second Part of the Popular Errors*, trans. Gregory David de Rocher (1579; reprint, Tuscaloosa: University of Alabama Press, 1995), 256–57. I here quote a physician on the subject of princely obligations because Joubert's account is particularly clear and extended, but his counsel is pervasively, if more diffusely, echoed in the practical ethical literature.

41. [William Cecil, Baron Burghley], *The Counsell of a Father to His Sonne, in Ten Severall Precepts, Left as a Legacy at His Death* (London, 1611), broadsheet; Dare, *Counsellor Manners*, 60–61.

42. Gainsford, *Rich Cabinet*, 36v.

43. Bryson, *From Courtesy to Civility*, 83–85; Brauer, *Education of a Gentleman*, 25–27; Thomas, "Health and Morality," 27; de Courtin, *Rules of Civility*, 123.

44. Peter Charron, *Of Wisdome*, trans. Sansom Lennard (London, 1612), 540; Burghley, *Counsell of a Father*.

45. Gainsford, *Rich Cabinet*, 51r.

46. Castiglione, *Courtier*, 299, 302.

47. Elyot, *Governor*, 209.

48. Ellis, *Gentile Sinner*, 131. See also the similarly Puritanical Richard Bra[i]thwait, *The English Gentleman. Containing Sundry Excellent Rules or Exquisite Observations, tending to Direction of Every Gentleman, of Selecter Ranke and Qualitie* (London, 1630), 305–72, esp. 306, 310.

49. Walter Ralegh, "Sir Walter Ralegh's Instructions to His Son and to Posterity," in *The Works of Sir Walter Ralegh, Kt.*, 8 vols. (London, 1876), 8:557–70, quotation on 568. Ralegh was here quoting Anacharsis, reputed to be one of the Seven Sages of Antiquity.

50. Lingard, *Letter of Advice*, 16–17.

51. William de Britaine, *Humane Prudence, or the Art by which a Man May Raise Himself & Fortune to Grandeur*, 3d ed. (London, 1686), 31.

52. Locke, *Some Thoughts Concerning Education*, 25.

53. Castiglione, *Courtier*, 302.

54. Elyot, *Governor*, 209.

55. Lodowick Bryskett, *A Discourse of Civill Life*, ed. Thomas E. Wright (1606; reprint, Northridge, Calif.: San Fernando Valley State College, 1970), 48, 162. This was apparently a popular early modern interpretation of Plato's *Republic*, IV, 430e–432b.

56. Gainsford, *Rich Cabinet*, 144r; Brathwait, *English Gentleman*, 311.

57. King James, *Basilicon Doron*, 100–101. See also Brathwait, *English Gentleman*, 305, 311. The identification of temperance as master virtue was not, of course, uncontested in the early modern period. For *prudence* "as the generall Queene, superintendent, and guide of all other vertues," see Charron, *Of Wisdome*, 350. Temperance was, for Charron, "not a speciall vertue, but generall and common, the seasoning sauce of all the rest" (532). The translator dedicated the book to Henry, Prince of Wales.

58. For the Galenic sources of the doctrine and phrase, see L. J. Rather, "The 'Six Things Non-Natural': A Note on the Origins and Fate of a Doctrine and a Phrase," *Clio Medica* 3 (1968): 337–47; Saul Jarcho, "Galen's Six Non-Naturals: A Bibliographic Note and Translation," *Bulletin of the History of Medicine* 44 (1970): 372–77; Peter Niebyl, "The Non-Naturals," ibid. 45 (1971): 486–92; Jerome J. Bylebyl, "Galen on the Non-Natural Causes of Variation in the Pulse," ibid. 45 (1971): 482–85; and Mikkeli, *Hygiene*, chap. 1. For a study of the nonnaturals in French early modern popular medicine, see Antoinette Emch-Dériaz, "The Non-Naturals Made Easy," in *Popularization of Medicine*, ed. Porter, 134–59, and Coleman, "Health and Hygiene in the *Encyclopédie*."

59. Henry Peacham, *The Truth of Our Times* (London, 1638), bound together with idem, *Complete Gentleman*, 175–239, quotation on 239.

60. Locke, *Some Thoughts Concerning Education*, 9–24. Among other practical ethical tracts organizing their advice—to a greater or lesser extent—through a list of the nonnaturals, see King James's *Basilicon Doron*.

61. For the medical context of Descartes's views on the passions, see, for example, Steven Shapin, "Descartes the Doctor: Rationalism and Its Therapies," *British Journal for the History of Science* 33 (2000): 131–54. For an entry into the early modern literature on the passions, see, for example, Susan James, *Passion and Action: The Emotions in Seventeenth-Century Philosophy* (Oxford: Clarendon Press, 1997); Stephen Gaukroger, ed., *The Soft Underbelly of Reason: The Passions in the Seventeenth Century* (London: Routledge, 1998); Cottingham, *Philosophy and the Good Life*, esp. chap. 3; and Jon Elster, *Alchemies of the Mind: Rationality and the Emotions* (Cambridge: Cambridge University Press, 1999).

62. George Cheyne, *An Essay of Health and Long Life* (London, 1724), 5. This passage is also quoted by Thomas's fine "Health and Morality," 24 (see 20–24 for Thomas's appreciation of the ethical significance of the nonnaturals).

63. Francis Bacon, "The History of Life and Death, . . . " in *The Philosophical Works of Francis Bacon*, ed. James Spedding, Robert Leslie Ellis, and Douglas Denon Heath, 5 vols. (London: Longman and Co., 1857–58; essay posthumously published 1636), 5:213–335, quotation on 299; also Steven Shapin, "Proverbial Economies: How an Understanding of Some Linguistic and Social Features of Common Sense Can Throw Light on More Prestigious Bodies of Knowledge, Science for Example," *Social Studies of Science* 31 (2001): 731–69, esp. 756–57.

64. In a mid-sixteenth-century translation of Erasmus's *Proverbs*, the Latin tag was given as "Without meate and drinke the lust of the body is colde"; alternatively, "The beste way to tame carnall lust, is to kepe abstinence of meates and drinkes"; and "A licorouse [licentious] mouth a licourouse taile." Desiderius Erasmus, *Proverbs or Adages*, ed. and trans. Richard Taverner (London, 1569), 34v. For medical proverbs generally, see Archer Taylor, *The Proverb* (1931; reprint, Berlin: Peter Lang, 1985), 121–29.

65. Cleland, *Instruction of a Young Noble-man*, 209. See also Gainsford, *Rich Cabinet*, 134v.

66. Peacham, *Complete Gentleman*, 151. See also Gainsford, *Rich Cabinet*, 36v; Castiglione, *Courtier*, 302. The literature produced by religious ascetics from the early Christian period is especially rich in appreciations of the causal influence of diet on sexual desire, while such ancient secular thinkers as Seneca laid much stress on the diet-anger connection. For an entry to this material, see Steven Shapin, "The Philosopher and the Chicken: On the Dietetics of Disembodied Knowledge," in *Science Incarnate: Historical Embodiments of Natural Knowledge*, ed. Christopher Lawrence and Steven Shapin (Chicago: University of Chicago Press, 1998), 21–50, and, especially, Bynum, *Holy Feast and Holy Fast*, 35–37.

67. Charron, *Of Wisdome*, 540.

68. George Cheyne, *The English Malady: or, a Treatise of Nervous Diseases of All Kinds*, ed. Roy Porter (1733; reprint, London: Tavistock/Routledge, 1991). See also Anita Guerrini, *Obesity and Depression in the Enlightenment: The Life and Times of George Cheyne* (Norman: University of Oklahoma Press, 2000), esp. chap. 6; Bryan S. Turner, "The Government of the Body: Medical Regimens and the Rationalization of Diet," *British Journal of Sociology* 33 (1982): 254–69; idem, "The Discourse of Diet," *Theory, Culture and Society* 1 (1982): 23–32.

69. I have written elsewhere about the practices and significance of philosophical asceticism, and their juxtaposition with early modern civic sensibilities: see Shapin, "Philosopher and the Chicken," esp. 33–37.

70. Robert Burton, *The Anatomy of Melancholy*, ed. Floyd Dell and Paul Jordan-Smith (1628; reprint, New York: Tudor Publishing, 1927), 200. The seventeenth-century sense of "Cockney" (a hen's egg, and by analogy a coddled child) did indeed pick out town dwellers in general and Londoners in particular but more directly pointed to people who were effete and squeamish—milksops. See also Piero Camporesi, *The Anatomy of the Senses: Natural Symbols in Medieval and Early Modern Italy*, trans. Allan Cameron (Cambridge: Polity Press, 1994), 65 (for hunger as "the cheapest and most universal of drugs"), and chap. 4 generally (for the early modern cultural significance of dietary abstinence).

71. Peacham, *Complete Gentleman*, 154. Peacham was here quoting *Ecclesiasticus*, a well-known compilation of maxims from the second century B.C.E.. The full passage is "What is life to a man derived of wine? / Was it not created to warm men's hearts? / Wine brings gaiety and high spirits, / if a man knows when to drink and when to stop; / but wine in excess makes for bitter feelings / and leads to offence and retaliation." Jesus Ben Sira, *Ecclesiasticus or The Wisdom of Jesus Son of Sirach*, ed. John G. Snaith (Cambridge: Cambridge University Press, 1974), 154.

72. See notably Mikhail Bakhtin, *Rabelais and His World*, trans. Hélène Iswolsky (1965; reprint, Bloomington: Indiana University Press, 1984). As Peter Burke nicely pointed out, "It was meat which put the *carne* in Carnival." Burke, *Popular Culture in Early Modern Europe* (London: Temple Smith, 1978), 186 (and chap. 7, for "The World of Carnival").

73. Francis Bacon, "The Advancement of Learning [Books I–II]," in *Philosophical Works*, 3:253–491, quotation on 373.

74. Francis Bacon, "Of Regiment of Health," in *The Essayes or Counsels, Civill and Morall*, ed. Michael Kiernan (1625; reprint, Cambridge: Harvard University Press, 1985), 100–102, quotation on 101. Bacon's was a creative reading of what Celsus actually had to say on the matter in *De medicina*, while the editor of this edition of the *Essayes* more circumspectly judges that "the notion of the 'benigne Extreme' is Bacon's emphasis" (237).

75. Bacon, "History of Life and Death," 261–62.

76. William Rawley, "The Life of the Right Honourable Francis Bacon," in *The Works of Francis Bacon*, ed. James Spedding, Robert Leslie Ellis, and Douglas Denon Heath, 15 vols. (Boston: Brown and Taggard, 1860–64; article comp. 1670), 1:3–18, quotation on 16–17.

77. Quoted in *Works of Francis Bacon*, 14:567.

78. Ibid., 14:261, 277, 295. Like Robert Burton (quoted above), Bacon was here explicitly criticizing the pedantic dietary precision of Luigi Cornaro's influential *De vita sobria* (Venice, 1558).

79. Bacon, "History of Life and Death," 304. This respecification of dietary moderation is briefly noted in Shapin, "Philosopher and the Chicken," 35–36. This is as far as Celsus went on the matter: "It is well . . . to attend at times a banquet, at times to hold aloof; to eat more than sufficient at one time, at another no more; to take food twice rather than once a day, and always as much as one wants provided one digests it." Aulus Cornelius Celsus, *De medicina*, trans. W. G. Spencer, 3 vols. (London: Heinemann; Cambridge: Harvard University Press, 1960), 1:43. But slightly later Celsus wrote, "Coming to food, a surfeit is never of service, [and] excessive abstinence is often unserviceable" (49).

80. Rawley, "Life of Bacon," 17 (for "grosser humours"); Bacon to Sir Humphrey May (May [1623]): "You may perhaps think me partial to Potycaries, that have been ever puddering in physic all my life." *Works of Francis Bacon*, 14:515. See also Lisa Jardine, *Ingenious Pursuits: Building the Scientific Revolution* (New York: Doubleday, Nan A. Talese, 1999), 294–95. For details of Bacon's self-medication, and also what he took from apothecaries, see "The Letters and the Life, Vol. IV," in *Works of Francis Bacon*, 11:28–29, 53–54, 78–80 (the *Memoriæ valetudinis*, recording Bacon's struggles with stone, gout, and melancholy); 13:200, 209, 328, 335; 14:10, 335, 398–99, 431, 514–15, 566–67 (his recipe for the rhubarb purgative and how he used it). On the mend from an illness, Bacon wrote to Buckingham (29 August 1623), "I thank God, I am prettily recovered; for I have lain at two wards, the one against my disease, the other against my physicians, who are strange creatures." *Works of Francis Bacon*, 14:431.

81. Michel Eyquem de Montaigne, "Of Experience," in *The Complete Essays of Montaigne*, trans. Donald M. Frame (1580, 1588; reprint, Stanford, Calif.: Stanford University Press, 1965), 815–57, quotation on 830.

82. Thomas Browne, *Pseudodoxia Epidemica: Or, Enquiries into Very Many Received Tenents, And Commonly Presumed Truths* (London, 1650), 229.

83. Laurent Joubert, *Popular Errors*, trans. Gregory David de Rocher (1579; reprint, Tuscaloosa: University of Alabama Press, 1989), 247; idem, *Second Part of Popular Errors*, 263.

84. John Aubrey, "Thomas Hobbes," in *Aubrey's Brief Lives*, ed. Oliver Lawson Dick (Ann Arbor: University of Michigan Press, 1975), 147–59, quotation on 155.

85. James Mackenzie, *The History of Health, and the Art of Preserving It*, 3d ed. (1758; reprint, Edinburgh: 1760), 125–26. Later on (135n.), Mackenzie made it clear that he was specifically criticizing Bacon's commendation of excess and the legitimacy of recruiting Celsus as an approving authority. See also Cheyne, *Essay of Health and Long Life*, 47–48. Others attributed the medical authority of such dangerous counsel to Avicenna.

86. Bacon, "Of Regiment of Health," 101. A historian of political thought perceptively discerns in this aphorism the "summation" of Bacon's moral philosophy: Ian Box, "Bacon's Moral Philosophy," in *The Cambridge Companion to Bacon*, ed. Markku Peltonen (Cambridge: Cambridge University Press, 1996), 260–82, on 278.

87. See, especially, Julian Martin, *Francis Bacon, the State, and the Reform of Natural Philosophy* (Cambridge: Cambridge University Press, 1992); also Stephen Gaukroger, *Francis Bacon and the Transformation of Early-Modern Philosophy* (Cambridge: Cambridge University Press, 2001).

88. La Rochefoucauld, *Maxims*, 155.

89. Rawley, "Life of Bacon," 17.

90. Celsus, *De medicina*, 1:43. See also Mark Grant, *Dieting for an Emperor: A Translation of Books 1 and 4 of Orabasius' 'Medical Compilations' with an Introduction and Commentary* (Leiden, The Netherlands: Brill, 1997), 12.

91. For example, Joubert, *Second Part of Popular Errors*, 263 (the healthy man "while he is feeling well, belongs to himself and does not have to follow any rule or diet nor consult a physician"); John Arbuthnot, *An Essay Concerning the Nature of Aliments, and the Choice of Them, according to the Different Constitutions of Human Bodies,* . . . 4th ed. (1731; reprint, London: J. and R. Tonson, 1756), 178–79 ("[A] healthy Man, under his own Government, ought not to tie himself to strict Rules, nor to abstain from any Sort of Food in common Use"); Mackenzie, *History of Health*, 135 ("A man who is sound and strong should ty himself down to no particular rule of diet, nor imagine that he stands in need of a physician"); Benito Jerónimo Feijóo y Montenegro, *Rules for Preserving Health, particularly with regard to Studious Persons*, trans. anon. from the Spanish (London: [1800?]), 79–80, 85. See also Mikkeli, *Hygiene*, 93–94, 102, 106, 108, 125, 130–31, 141. For the Rule in a popular medical text by a nonprofessional, see Elyot, *Castel of Helthe*, 45r ("[A]s Cornelius Celsus saith, A man that is hole and well at ease, and is at his lybertie, ought not to bynde him selfe to rules, or nede a phisition"); also Leonard Lessius, *Hygiasticon: Or, A Treatise of the Means of Health and Long Life*, in *A Treatise of Health and Long Life, with the Sure Means of Attaining It, in Two Books. The First by Leonard Lessius, The Second by Lewis Cornaro, a Noble Venetian: Translated into English, by Timothy Smith, Apothecary* (ca. 1600; reprint, London, 1743), 23–24 ("For [generally speaking] any Sort of Food that is common to one suits agreeably enough with hale Constitutions").

92. King James, *Basilicon Doron*, 125, 127; and, following the king, Cleland, *Instruction of a Young Noble-man*, 212: "I thinke it best to accustome your selfe unto the Countrie where you are."

93. De Courtin, *Rules of Civility*, 128–30.

94. Peacham, *Complete Gentleman*, 151. See also Lingard, *Letter of Advice*, 41.

95. Locke, *Some Thoughts Concerning Education*, 11–12.

96. Gailhard, *Compleat Gentleman*, pt. 1, 85–86; Edward Panton, *Speculum Juventutis: or, a True Mirror* . . . (London, 1671), 188.

97. Montaigne, "Of Experience," 821.

98. Descartes's version ran this way: "So, as Tiberius Caesar said (or Cato, I think), no one who has reached the age of thirty should need a doctor, since at that age he is quite able to know himself through experience what is good or bad for him, and so be his own doctor." John Cottingham, ed. and trans., *Descartes' Conversations with Burman* (Oxford: Clarendon Press, 1976), 51. The source is in fact Suetonius's life of Tiberius.

99. Montaigne, "Of Experience," 827, 830, 832. See also Margaret Brunyate, "Montaigne and Medicine," in *Montaigne and His Age*, ed. Keith Cameron (Exeter: University of Exeter Press, 1981), 27–38.

100. Montaigne, "Of Experience," 827.

101. Ibid., 833.

102. Ibid., 832.

103. For a summary of the links between free action and gentlemanly identity in early modern English culture, see Steven Shapin, *A Social History of Truth: Civility and Science in Seventeenth-Century England* (Chicago: University of Chicago Press, 1994), esp. chaps. 2–3.

104. Montaigne, "Of Experience," 830–31. Compare Charron: "It is one of the vanities & follies of man, to prescribe lawes and rules that exceed the use and capacitie of men, as some Philosophers and Doctors have done. They propose strange and elevated formes or images of life, or at leastwise so difficult and austere, that the practice of them is impossible at least for a long time, yea the attempt is dangerous to manie." Charron, *Of Wisdome*, 197.

105. Montaigne, "Of Experience," 835.

106. Joubert, *Second Part of Popular Errors*, 262.

107. Lessius, *Hygiasticon*, 2–3. Lessius was not himself a physician; he taught philosophy and divinity at the Jesuits' college at Louvain.

108. Cheyne, *Essay of Health and Long Life*, 4.

109. On these points, see the classic papers by N. D. Jewson, "Medical Knowledge and the Patronage System in Eighteenth-Century England," *Sociology* 8 (1974): 369–85, and "The Disappearance of the Sick-Man from Medical Cosmology, 1770–1870," ibid. 10 (1976): 225–44.

110. For example, Charles E. Rosenberg, "Banishing Risk: Continuity and Change in the Moral Management of Disease," in *Morality and Health*, ed. Brandt and Rozin, 35–51.

111. Julia Child, quoted in Calvin Tomkins, "Table Talk," *New Yorker*, 8 November 1999, 32.

MARY E. FISSELL

3

Making a Masterpiece

The Aristotle Texts in Vernacular Medical Culture

Popular medical books, those small works explicitly intended for patients rather than practitioners, have left few historical traces. Hundreds of different texts from the seventeenth and eighteenth centuries survive today, usually in copies that look well worn. We presume that some were publishing successes because they went into many editions. Nicholas Culpeper's *English Physician,* first published in 1652, was republished at least fifty-four times in the next century and a half, and many times thereafter, often under the title *Culpeper's Herbal.* Eighteenth-century bestsellers included William Buchan's *Domestic Medicine* (at least fifty-two editions to 1800) and John Wesley's *Primitive Physick* (at least twenty-four editions to 1800).[1] Despite such apparent success, however, these books have left few historical traces, and we have little evidence about how they were used. Those standbys of social history, letters and diaries, do not often mention such works. Nor do probate inventories list many popular medical works because often they were worth too little to mention.

One book, however, provides something of an exception to this historical silence. This is *Aristotle's Masterpiece,* an amalgam of midwifery advice spiced with a few hints about sexual intercourse.[2] References to *Aristotle's Masterpiece,* first published in 1684, appear in a range of sources. For example, the English radical Francis Place read *Aristotle's Masterpiece* as a schoolboy in the late eighteenth century. As he explained in his autobiography, "This I contrived to borrow and compared parts of it with the accounts of the Miraculous Conception in Matthew and Luke, and the result was that in spite of every effort I could make I could not believe the story."[3] Place's account reminds us that readers may take very different messages from a text than the text seems to intend. The introduction to the *Masterpiece* stresses its connections to biblical authority rather than suggesting that the natural knowledge it purveys might undermine scriptural accounts. Place ignored the ostensible message, taking a fragment of the text and bending it to his own purposes. Historians of early modern reading practices have shown us telling examples of such unorthodox or unintended readings, the best known of which landed a reader in front of the Inquisition.[4] As we shall see, Place was not the only boy to peruse the *Masterpiece*, which claimed to be for married women and men.

In what follows, I explore the traces that the *Masterpiece* has left in the historical record. These traces suggest how important cheap printed books were to ordinary people's perceptions of their own bodies. The meanings ascribed to such books are complex and often unexpected. These little books were read and used in ways apparently unforeseen by their writers and publishers, or at least in ways unacknowledged by them. The multiple uses of the *Masterpiece* show us the extent to which popular medical books were both a part of a larger world of cheap print and integral to vernacular cultural productions of human bodies.

While the book is singular in its popularity and longevity, some of the uses to which it was put may have larger relevance to our understanding of popular medical works. The richness of the traces left by this work show us how ordinary people's understandings of their bodies might have been deeply influenced by cheap print. Books and bodies, I argue, existed in a kind of reciprocal relationship. Ideas about how the body worked were influenced by popular medical books (whether consumed directly or indirectly), but at the same time, successful books tended to offer ideas that were congruent with their readers' expectations and beliefs. In the case of the *Masterpiece*, two patterns of circulation suggest why the book was so outstandingly popular. First, the work circulated between realms of knowledge presumed to be the purview of women and those more often associated with men. Second, although a printed book, the work circulated in spoken words as well as written ones.

The *Masterpiece* and three other "Aristotle" texts often published with it dominated the market for popular medical books in the eighteenth century. From the 1690s to 1800, over half of the editions of all popular English books about reproduction were Aristotle texts.[5] No other single text or group of texts came anywhere near such renown. While the book achieved market domination in the eighteenth century, it continued to be a bestseller for centuries. Countless nineteenth-century editions, often undated, continue to occupy the shelves of rare-book sellers and collectors. In the early twentieth century, James Joyce referred to the work as a commonplace in *Ulysses*. Leopold Bloom looks at a copy of the *Masterpiece* at a Dublin bookstall (he has already given Molly Bloom her own copy). She calls it "Aristocrats Masterpiece."[6] When the noted Shakespeare scholar A. L. Rowse went up to Oxford in the 1920s, he never mentioned to his mother that he was studying Aristotle: "Aristotle would have meant to my mother, as secretly to Victorian women, his book on child-bearing: unmentionable. But I knew that that book was secreted in her chest of drawers in the old home."[7]

The copy that Mrs. Rowse kept tucked away in her chest of drawers was a slightly modified version of the edition of the *Masterpiece* that first appeared in 1702. The medical historian Sir D'Arcy Power recalled slinking into a shop selling contraceptives in London in the late 1920s and buying a recent edition of the *Masterpiece* there.[8] Nineteenth- and early-twentieth-century editions had illustrations different from those in earlier editions, but much of the text remained the same for hundreds of years. The book was published as late as the 1920s.

History of the Text

The *Masterpiece*'s history stretches back to the late sixteenth century as well as forward into the twentieth. While the book itself first appeared in 1684, its roots went deeper. The work was often bound with three other pseudo-Aristotle texts, or in the latter half of the eighteenth century, combined into a book called *The Works of Aristotle*. Of these four, the oldest is *The Problemes of Aristotle* or *Aristotles Problems*, first published in English in 1595.[9] The text is a series of questions and answers, gathered under various headings. Questions about sex and reproduction are included, although the text addresses many aspects of human and animal function. The *Problems* was a steady favorite with English readers, reaching at least six editions by the period of the English Civil War, and another ten by 1700. In the first half of the eighteenth century, numerous "twenty-fifth" editions were published.

The next Aristotle text to appear was *Aristotle's Masterpiece*, extant in at least three different versions. The first was published in 1684; it addresses sex, conception, pregnancy, and childbirth. The second version, first published in 1697, is a combination of the first version and another medical work on the diseases of women. These two versions were slowly superseded by the third one, appearing first in 1702, which is a rewritten and recombined version of the second. It is this third version that lived on well into the nineteenth century. I shall refer to these various versions as *Masterpiece 1, Masterpiece 2,* and *Masterpiece 3.*[10]

At about the same time that *Masterpiece 2* was being produced, a third Aristotle text was published. Called *Aristotle's Legacy* or *Aristotle's Last Legacy,* this text is also extant in at least two very different versions. Initially, the book had sections on fortunetelling, palmistry, the interpretation of dreams, and the meaning of moles and blemishes on the face. This part of the text resembles books of knowledge, such as the *Erra Pater* or the *Compost of Ptolemeus.* Unlike most of those works, the *Legacy* includes a selection of riddles and jokes, and a canting dictionary. By the 1730s, this *Legacy 1* was superseded by *Legacy 2,* a rewritten version of the *Masterpiece* addressing conception and childbirth.[11]

In 1700, a book called *Aristotle's Compleat and Experienc'd Midwife* was published. This work consists of revised chapters on conception, childbirth, and lying-in from *Masterpiece 1* and from Nicholas Culpeper's *Directory for Midwives,* first published in 1651. It went on to be published another twenty times over the course of the eighteenth century. Unfortunately, most editions lack both a date of publication and any indication of who published it.[12]

For a brief moment in the 1690s and 1700s, the name "Aristotle" seems to have been especially favored. It is as if the initial popularity of the *Masterpiece* prompted publishers and printers to try out a number of "Aristotle" texts. For instance, another version of the *Problems,* called *Aristotle's New Book of Problems,* was first issued in this period.[13] *Aristotle's Manual of Choice Secrets* was issued once, in 1699.[14] It is a greatly abbreviated version of Jacques Guillemeau's midwifery text, first published in English in 1612. It is so abbreviated that parts of it contradict its source. In 1705, an enterprising bookseller published *A Family Jewel,* a hybrid of the *Masterpiece* and the *Erra Pater,* a book of knowledge.[15] Another edition of the work was listed in the 1704 Term Catalogue as *Aristotle's Family Jewel,* although I have not found reference to any extant copy. After about 1710, this flurry of other texts ceased, and the market settled down to what had become the four customary Aristotle texts.

The *Masterpiece,* like other texts of its kind, was not newly created in 1684. Instead, it is made up of pieces of older works. The first version is largely made up of two texts: *The Compleat Midwifes Practice,* and Levinus Lemnius's *The Se-*

cret Miracles of Nature.[16] *The Secret Miracles of Nature* is a vast compendium of natural knowledge, addressing reproduction, disease, and the functioning of the natural world and offering numerous precepts for Christian conduct.[17] Its table of contents alone runs to eight tightly printed, double-columned large pages.

However, I suspect that the compiler of the *Masterpiece* actually used a text called *A Discourse Touching Generation*, extracts of the larger Lemnius text published in 1664, and again in 1667.[18] This tiny book includes the first sixteen chapters of the larger text, and then selects chapters on barrenness, venereal disease, the connection between gout and lasciviousness, and another five chapters on sex and marriage. Everything of Lemnius's text included in the *Masterpiece* is in the *Discourse*. Of the twenty-four *Discourse* chapters, thirteen are partly or wholly included in the *Masterpiece*.

The second version of the *Masterpiece*, first published in 1697, is called *Aristotle's Master-piece Compleated*. The two-part structure referred to in the title is a combination of *Masterpiece 1* and a 1636 book by John Sadler, a Norwich physician, called *The Sick Woman's Private Looking-Glasse*. Sadler's text focuses not on generation but on the various ills to which a woman's reproductive system is prone.[19] *Masterpiece 2* thus includes chapters taken from Sadler on topics such as the retention of the menses, the whites (a vaginal discharge), inflammation of the womb, and the falling-down of the womb. Missing from this second version are a number of chapters on problems associated with the period of lying-in after childbirth, which had been part of *Masterpiece 1*. The second version was fairly short-lived; by the 1710s it too was superseded by the next version of the text, although, of course, many copies of the earlier versions continued to be read, and some late-century provincial editions resurrected it as well.

The third version of the *Masterpiece* is substantively different from the first two versions. Some pieces of it derive from the two earlier versions. Other parts are taken from Nicholas Culpeper's *Directory for Midwives* and John Pechey's version of the *Compleat Midwife's Practice Enlarged*.[20] Some parts, such as the section on virginity, appear to be new discussions of topics covered in the earlier versions.[21] The range of topics—conception, fetal development, pregnancy, labor and delivery—corresponds closely with earlier versions. However, the third version includes two new sections, one on physiognomy, the other entitled "The Family Physician," a series of recipes for domestic remedies. The introduction claims that the recipes come from the great Hippocrates himself.

Sexual Knowledge and the Spread of Inexpensive Books

While the bibliographic history of these pseudo-Aristotle texts is complex, it is only one element in my analysis of the cultural functions and meanings of these works. For "Aristotle" and his knowledge of the human body did not live just between the covers of books. The many ways in which Aristotle was invoked in early modern England and colonial America illuminate the deep interdependence between vernacular body cultures and cheap medical print. Historians have shown that the dichotomy of print versus oral culture does not adequately describe political activism or literary production in seventeenth-century England. Nor does that dichotomy describe the relations between bodies and books about bodies.[22] In what follows, I explore a number of moments in which Aristotle and the books that bore his name were used in a variety of contexts in relation to the body. These moments show us how much bodies and cheap books were intertwined, for readers and nonreaders alike.

Aristotle's Problems was first published in English in 1595. Almost immediately, the name "Aristotle" became associated with sexual knowledge—or perhaps Aristotle, the greatest classical authority on generation, was already associated with sex in some circles. Thomas Middleton's 1602 play *The Family of Love* includes a scene in which a doctor's wife diagnoses pregnancy in another woman, saying, "Aristotle speaks English enough to tell me these secrets."[23] Here the relationships among classical authority, learned print culture, and women's knowledge about female bodies are recombined in an unexpected and comical way. Aristotle's Greek becomes English, and metaphorically, the doctor's wife reads both her patient's body and a classical authority in her attempt to determine pregnancy.

This female character appropriates classical learning to a very different tradition of women's secrets. Before the advent of modern diagnostic techniques, it was not easy to determine pregnancy in the early months. The clearest indication was quickening, when the mother felt the baby move inside her, but quickening did not happen until the fourth or fifth month of pregnancy. Therefore, vernacular medical works as well as women's domestic practices employed a range of methods, from urine tests to a panoply of subtle physical signs, to establish pregnancy.[24] Women's knowledge of the female body's reproductive capacities was widely respected; juries of matrons, for example, determined if a woman condemned to death for a crime was truly "pleading the belly," claiming pregnancy as a way to postpone execution.[25] In cases of suspected infanticide, too, women often were called upon (or nominated themselves) to inspect the body of a woman under suspicion, reading its signs with care for evidence of delivery.[26] Knowledge of the female body's reproductive aspects was considered women's expertise, and women were granted social authority to deploy that expertise.

Two years after Middleton's *Family of Love*, Thomas Dekker's play *The Honest Whore* also includes a character who diagnoses pregnancy having read "Albertus Magnus, and Aristotles emblemes."[27] "Albertus Magnus" refers to a medieval text, spuriously attributed to Albertus Magnus, usually titled *The Secrets of Women*.[28] I have never found any reference to a text called "Aristotle's Emblems." In all likelihood, this too is an allusion to the book of his problems, which had been published recently. "Emblem" and "problem" sound alike; the substitution may have been meant as a joke at the expense of the female character who cites learned authorities. Each also refers to a text made up of many little self-contained entries, either emblems or individual problems.

"Aristotle" continued to be invoked as a source of reproductive and sexual knowledge throughout the seventeenth and eighteenth centuries. In Edward Ravenscroft's 1681 play *A London Cuckold*, Alderman Wiseacre explains his choice of a fourteen-year-old bride, complaining: "Girles now at sixteen are as knowing as Matrons were formerly at sixty, I tell you in these days they understand Aristotle's Problems at twelve years of age."[29] In all three of these examples from plays, we see how the figure of Aristotle was closely connected to sexual or reproductive knowledge, especially the knowledge usually considered women's property rather than men's.

The connections between the *Legacy* and the London stage remind us that cheap print and oral performances were very closely linked. While historians have largely abandoned any rigid dichotomies between oral and print cultures, the circulation of the figure of Aristotle can help us to understand how even some unlikely texts were read aloud, consumed by hearers as well as readers. The *Problems* is itself a text that refers to oral rather than printed forms of knowledge production and dissemination. The book consists of questions and answers, modeled upon scholastic forms of pedagogy that emphasized oral performance rather than command of written language.[30] For example, the text asks about some of the confusion surrounding the early diagnosis of pregnancy when it poses the question, "Why do they [the menses] run the first three moneths [*sic*] in women with childe?" to which the reply is, "By reason of the smalnes of the child, which cannot take all that matter and substance."[31] It is a historical irony that devices such as the question and answer format and divisions into books and chapters, originally used in a learned manuscript culture in which written texts were a scarce resource, enabled early modern novice readers to navigate the unfamiliar world of the printed text. Thus, both in the ways in which the book is invoked by characters and in its own composition, the *Problems* circulates between spoken and written language, and between book and body.

Schoolgirls are also depicted as readers of the *Problems* by Isaac Bickerstaffe, the epistolary voice of the *Tatler*. Supposedly penned at Will's Coffee-House, or the Grecian Coffee-House, as well as from Bickerstaffe's own home, these letters attempt to create a world peopled by the polite. Bickerstaffe set up a competition to nominate the twelve most important figures for his Chamber of Fame. Like *Time* magazine's Man of the Year, Bickerstaffe's individuals were to be chosen on the basis of fame rather than virtue. However, Bickerstaffe encountered an unexpected problem: what to do with men to whom fame is falsely attributed? He had asked his sister Jenny about her nominations for the chamber, mentioning Aristotle in particular. "She immediately told me, he was a very great Scholar, and that she had read him at the Boarding-School. She certainly meant a Trifle sold by the hawkers, call'd *Aristotle's Problems*."[32] Bickerstaffe, having made a joke at his sister's expense, implies that his (male) readers will have some acquaintance with the "real" Aristotle.

Bickerstaffe is at pains elsewhere to paint Jenny as an honest and moral woman. But his very insistence calls female virtue into question. He has promised to get his sister "an agreeable Man" for a husband if she keeps "her Honour," and he emphasizes that she is as "unspotted a spinster as any in Great Britain."[33] His concerns echo other contemporary images that doubt women's sexual virtue. In the early eighteenth century, girls in the new boarding schools were the subject of lurid depictions of the dangers of the female imagination driven by sexual curiosity. The tenth edition of *Onania* (1724) tells a story in a letter supposedly sent in by a reader. When the writer was at boarding school at the age of fifteen, she was "shewed the way" by three of her friends: "I followed it afterwards upon all Opportunities by my self, and so by the Practice, and the lascivious talk we had amongst ourselves, and Play Books, and other Books we us'd to read one to another . . . I was resolved to marry the first man that ask'd me the Question."[34] Just as in many eighteenth-century novels, dire consequences ensue when a young woman rushes into an ill-advised marriage.

In this letter, the young woman's hasty marriage is followed by widowhood and barrenness. In this story, too, printed words become spoken words. The girls read their lascivious books out loud, unsupervised by parents. Like the plays mentioned in this instance, *Aristotle's Problems* was also particularly suited to reading aloud because of its dialogue format.

This scene, probably imagined by the author of *Onania* rather than experienced by a genuine letter writer, suggests the ways in which knowledge of reproductive bodies was understood to be female property. The group of girls is almost a parody of the group of women who attended childbirth. The social practices surrounding childbirth emphasized the exclusion of men from women's knowledge. Well before a woman went into labor, she extended invitations to her closest female relatives, friends, and neighbors to support her through the birth. As many as a dozen women stayed with the laboring woman, trying to keep up her spirits by talking, telling jokes, and drinking a special kind of wine thickened with grain. These hours of labor were dangerous and frightening, but the men excluded from the birthing room imagined the women telling jokes about sex, belittling men's performances and reputations. The women invited to the birth were called "gossips," from which our modern use of the word "gossip" to mean scandalous and intimate personal details is derived. The word "gossips" was also, by extension, used to describe the visitors to the new mother, who expected to be feasted to celebrate the new arrival.[35]

Two further examples suggest the ways in which ordinary people's ideas about their bodies were shaped by the interplay between spoken and written words. In each case, the uses to which *Aristotle's Masterpiece* was put remind us that books are not always read in the ways that their authors claim they should be. Perhaps the *Masterpiece*'s remarkable success in the eighteenth century was due, in part, to the ease with which it could be adapted for multiple purposes. Nor can the *Masterpiece* be understood as a closed text, fixed in print. Rather, its readers brought their own, extratextual knowledge and, in effect, inserted it into the book. As we shall see, the book continued to function as an icon of sexual knowledge, just as the bantering stage references to *Aristotle's Problems* had done for an earlier pseudo-Aristotle text.

John Cannon's memoirs give us a rare snapshot of one person's use of the *Masterpiece*. Born in Somerset in 1684, Cannon grew up in an agrarian community until at the age of twenty-three he took up a commission as an excise officer, traveling from place to place. As Tim Hitchcock has shown, Cannon's

sexual behavior was strongly patterned after the norms of his immediate communities.[36] As a farm laborer in West Lydford, Cannon courted a young woman
and committed himself to her in the traditional ritual of breaking a shilling,
each partner carrying half. Once he was on the road, however, Cannon's behavior was much more strongly influenced by other young men who had also
left customary constraints behind in their rural villages. Cannon was a love-
them-and-leave-them traveler, almost a parody of the type.

Cannon wrote his memoirs later in life, and he gives us a rich and sometimes startling picture of many aspects of provincial culture. He describes in
detail, for example, how at the age of twelve he and a group of other boys were
taught to masturbate by a seventeen-year-old when they went for an after-
school swim.[37] Cannon attended school until the age of thirteen, where he
learned to read and write and cast up accounts, as well as the rudiments of
Latin. He borrowed books from friends and was an active reader. He purloined
his mother's copies of Nicholas Culpeper's *Directory for Midwives* and *Aristotle's
Masterpiece*, noting in his memoirs that he wanted to know what women
looked like underneath their clothes.[38] Indeed, he adopted the very phrase used
in the title of the *Masterpiece:* he wanted to know about the "secrets of nature."
"Secrets" was a very sexually loaded term for occult knowledge because it also
meant the female genitals. Later, when living the roisterous life of an excise-
man, Cannon uses the word in describing an assault upon a maidservant: another young man, Cannon writes, "laid his hand upon her secrets which she
being sensible of began to curse."[39]

The Gendered Audience: Titillation versus Modesty

Cannon's choice of the *Masterpiece* was, needless to say, constrained by his
mother's collection of books on midwifery. But his interest in what women
looked like underneath their clothes was addressed by the sexual language of
"the secrets of nature" and by the frontispiece to the *Masterpiece*. While we cannot know for certain which version or edition Mrs. Cannon owned, it is most
likely that her copy included a woodcut frontispiece of a hairy and naked
woman.[40] Although the image was intended in part to warn the reader about
the powers of the female imagination (her mother had gazed upon a picture of
St. John the Baptist wearing animal skins while she was pregnant and thus
marked the unborn child), it was also one of the few images of a naked female
body in cheap print. While there were a few biblical images of Adam and Eve,

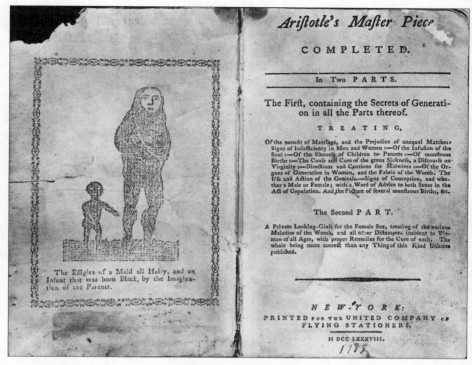

Aristotle's Master Piece Completed (New York: Printed for the United Company of Flying Stationers, 1788). The frontispiece to the 1788 *Masterpiece*. Earlier versions of this image emphasized the woman's nudity, providing a titillating spectacle for a potential reader of the book. Courtesy of the Library Company of Philadelphia.

or of Susannah bathing while the elders watched, such pictures were rare. There were also the intentionally scandalous images associated with the radical sects of the English Civil War, where religious gatherings were portrayed as little better than orgies.[41] However, the *Masterpiece* image was one of the most easily accessible pictures of a naked woman in cheap print.

When Cannon's mother caught him masturbating with her midwifery books, she promptly repossessed them, whereupon Cannon pursued his quest for knowledge of the female body by drilling a small hole in the wall of the family privy, through which he spied upon one of the family's maidservants. He employed this peephole for years, often masturbating while spying. Earlier historians of popular medical print, such as Peter Wagner, have suggested that such small books were intended to titillate—that Cannon's use of them was perhaps that envisioned by printers and publishers.[42] However, I would emphasize the multiple uses of such texts. After all, these are Mrs. Cannon's books, and she is the stated intended reader for them. The title page of the first version of the *Masterpiece* proclaims the book to be "very necessary for all Midwives, Nurses, and Young-Married Women." Mrs. Cannon's other midwifery book, Culpeper's *Directory*, was subtitled "A Guide for Women."

Many of the themes about the relationships between male and female knowledges of the body and between written and spoken versions of that knowledge reappear in the best-known incident involving *Aristotle's Masterpiece*. In 1744, the minister Jonathan Edwards learned that young people in his Northampton, Massachusetts, parish had been reading a book on midwifery and consequently indulging in "lascivious and obscene Discourse."[43] Edwards investigated, drawing up lists of young men who had read the book and young women who could testify that the young men teased them about sexual matters with reference to the book. Although initially approving of Edwards's investigation, the congregation then turned against it. This so-called bad book controversy has been discussed by a number of historians of New England, from the eighteenth century onwards, sometimes as a harbinger of Edwards's future difficulties with his parish.[44]

As we shall see, the "bad book" was actually three different books, one of which was almost certainly *Aristotle's Masterpiece*. By 1744, the *Masterpiece* was the single most important popular book about reproduction. Twenty-three editions had been published since its first appearance in 1684, as had more than fifty editions of the other three pseudo-Aristotle texts related to it. The *Masterpiece* was a North American favorite in particular, although, like many others, the book was not actually published in America until after the revolution. A handful of early London editions were published by Benjamin Harris, who fled to Boston in the 1680s to avoid the consequences of his religious and political advocacy.[45] Harris may well have sold copies of the *Masterpiece* during his sojourn in Boston. By the early nineteenth century, the book was so popular in America that when Parson Weems, traveling salesman for the Philadelphia publisher Matthew Carey, wanted to unload a dull midwifery text, he called it the "Grand American Aristotle." Then it sold, he wrote, "like green peas in spring."[46] Edwards's detailed notes about the controversy in his parish provide us with a wealth of information about the local uses of two or three midwifery books. Medad Lyman and John Lancton each referred to a book they called "Aristotle." Lyman wanted to find the book, while Lancton boasted that he had read it more than once.[47]

In the 1930s, Thomas Johnson first suggested that the "Aristotle" to which these young men referred was one of the four pseudo-Aristotle texts gathered together as *Aristotle's Works* as early as 1733. While Johnson is correct, Edwards's notes provide further clues that can help us identify the text as the *Masterpiece* in particular. Inadvertently, Johnson and other historians have conflated two or three different midwifery books that circulated in Northampton. Elizabeth Pomeroy and Katherine Wright testified that they had found a book in Pomeroy's house, tucked up in the backside of a chimney, called *The Midwife Rightly Instructed*. Edwards wrote "A new book" in his notes on Pomeroy's testimony, and he was correct about the book's "newness" in at least two ways. First, this book was new because it was not the same as that referred to by Lyman, Lancton, and a number of other young men. In all likelihood, Pomeroy and Wright looked at a copy of Thomas Dawkes's *The Midwife Rightly Instructed*.[48] Since the book had been published only eight years earlier in London, Edwards, or his witnesses, also quite rightly considered this a "new" book in the sense that it had just been published.

Edwards's young parishioners also mentioned a third midwifery book in Northampton. Rebeccah Strong told Edwards that five years earlier, Charles Wright and Timothy Root had come into her shop, with the book "that they were provoked about." When asked if she had ever seen such a book before, Strong testified that she had, at Dr. Samuel Mather's house. She concluded from a glance at the pictures "that seemed to her to be parts of a woman's body" that the book Wright and Root brandished was the same as or similar to that owned by Mather.[49] We will probably never know what book Mather owned, but Strong's comment about the picture makes identification of the "bad book" more certain. Of the four pseudo-Aristotle texts available in 1744, only one had an image that would have been described as parts of a woman's body. The third version of the *Masterpiece* included a folding leaf plate of "The Form of a Child in the Womb" in addition to a startling frontispiece derived from that in the first version. In both the anatomical image and the frontispieces, female genitals and breasts are clearly depicted. None of the other three pseudo-Aristotle texts (the *Compleat and Experienc'd Midwife*, the *Problems*, or the *Last Legacy*) included this sort of anatomical image.

Indeed, such images were comparatively rare in popular midwifery books.

Sometimes, such books might include the so-called baby in bottle illustrations depicting various ways in which an infant might present poorly for a normal labor and delivery. Woodcuts cost money. Even if a printer already owned a suitable block, its inclusion in a book always disrupted the flow of printing. Many early modern popular midwifery texts did without anatomical images. Some editions of Nicholas Culpeper's best-selling *Directory for Midwives* and various versions of the *Compleat Midwifes Practice Enlarg'd*, for example, included woodcut frontispieces of their authors but lacked anatomical plates. Such an omission was not merely due to cheapness on the part of a printer. Midwives emphasized touching—knowing how a baby presented by manual examination—rather than the more surgically oriented understanding promoted by an anatomical image.[50]

Dawkes addresses his book both to midwives and to married women, claiming that English midwives are very ignorant and need instruction. For further reading, he recommends works by Deventer, Giffard, and Maubray. Only the last of these—John Maubray's *The Female Physician*—can be considered "popular" in the sense of being addressed to mothers as well as to male and female midwives.[51] William Giffard was the first male midwife outside the Chamberlen circle to advocate for the use of forceps, but Dawkes himself only alludes to the use of the fillet for difficult births.[52] Dawkes can be read as a text that disciplines midwives. It is in the form of a dialogue between a surgeon and "Lucina," a midwife. The surgeon is always the one offering knowledge, and Lucina's lines consist of inquiries and grateful thanks for information received. The book contains no pictures; Dawkes explains that such images should stay in surgeons' museums and not be shown to a larger public. While Dawkes believes that midwives should have anatomical knowledge, he wants the midwives to access that information via surgeons. When the surgeon shows Lucina a skeleton so that she can understand the structure of the pelvis and how it functions in delivery, Dawkes has the two retire to the surgeon's study, and the reader never learns what transpires within it. Dawkes keeps anatomical knowledge hidden from his readers.

While we might read Dawkes's coyness about anatomical knowledge as, in part, a move to ensure a role for surgeons in the delivery of difficult cases, he emphasizes the requirements of female modesty. In his preface, he claims, "And the ladies may not think, that, by offering the Tract itself to their perusal, I offer an indignity to their Chastity; I here assure them, that I have, throughout the whole discourse, inviolably preserve'd it, from any expression which wou'd clash with the principle of modesty, which is peculiar to their sex; there being not a single sentence in it (the nature of the subject considered) which might put the chastest of them to the blush."[53] Dawkes is part of a larger move to make the knowledge of women's bodies fit for polite consumption. One of the books he recommends, Maubray's *The Female Physician,* is similarly engaged in a project of politeness. Maubray draws extensively upon Levinus Lemnius, the same writer whose work is incorporated in the *Masterpiece.* However, Maubray clothes his Lemnius in elaborate and flowery prose, and dedicates the work to "all Learned and Judicious Professors of Physick, as well as Ingenuous and Experienced Practisers of Midwifery." When Betty Pomeroy and Katherine Wright find a copy of Dawkes tucked up in a chimney shelf, in other words, they find a book with a very different attitude towards knowledge of the body than that of the *Masterpiece.* While Dawkes's work points towards the overlap between spoken and written words since it is in dialogue form, it denies that women have knowledge of female bodies. Between male and female knowledge of the body there is segregation and prohibition rather than circulation.

While Dawkes is concerned with the modesty of his female readers, worry is expressed in the *Masterpiece* about improper male readers, a trope common to popular midwifery books since their first publication in English in the sixteenth century. As the examples of John Cannon and the young men in Edwards's Northampton parish suggest, concern about male readers was not solely a ritual bow towards propriety or a tantalizing come-on, but also evidence of deeper attitudes about the knowledge of female bodies and the power it might convey to those who should not have it.

In the earliest printed midwifery texts in English, all of which were translations from continental sources, authors fret about the possibility that young men might read their work. Eucharius Roeslin, the author of the most important pre-1650 popular book about midwifery published in England, worries about communicating the knowledge of women's bodies to men. In his preface to readers, Roeslin considers the possibility that men who read the book or hear it read aloud "shalbe moved thereby the more to abhor and loath the company of Women" or "to jest and bourd of Women privities."[54] Roeslin hastens to defend his book, noting that the wonders of the body can promote godly devotion, and adds that any item can be used for good or evil. He goes on to warn his readers that anyone who uses his book to speak irreverently of women's bodies does "great injure, dishonor, and contumely to Nature" and is guilty of "mortal and deadly sinne."[55] Almost half of Roeslin's introduction is taken up with concerns about his book being read by lascivious or misguided men.

James Guillemeau, whose French midwifery text was intended in part for young male surgeons, was translated into English by a writer concerned with the potential impropriety of writing about women's bodies. The translator expresses concern about having "been offensive to Women, in prostituting and divulging that, which they would not have come to open light, and which beside cannot be exprest in such modest termes, as are fit for the virginitie of pen & paper, and the white sheetes of their Child-bed."[56] This image, in which ink on white paper is metaphorically transposed into virginal blood on white sheets, suggests some of the multiple anxieties men had in talking about women's private parts in print. The translator defends himself by saying that he has tried to be "private and retired" in the way he writes about women's bodies.[57]

In the introduction to the third version of the *Masterpiece,* similar concern is voiced about the propriety of writing about women's bodies: "[T]his Knowledge is too often abus'd by vain and light Persons, who instead of admiring the Wisdom of God in the secrets of Generation, do only make it their Business to ridicule and set 'em at nought."[58] This version of the *Masterpiece* repeatedly emphasizes that knowledge about sex is acceptable because of its divine origin. The author of the preface points to Job (the passage in which human generation is compared to milk curdling to make cheese, Job 10:10) and to David (Psalm 138, in which a fetus is described as "fearfully and wonderfully made") as models for their inquiries into generation. Verses of the same psalm are printed on the anatomical plate in the text. At the same time, however, this text uses sexually charged metaphors of penetration to describe the acquisition of such knowledge. On the very first page of the first chapter of the book, knowledge of reproduction is "only know[n] to those who trace the secret Meanders of Nature in their private Chambers."

Further evidence suggests that the *Masterpiece* invited multiple readings. While the introduction cautions readers about the abuse of sexual or reproductive knowledge, elsewhere the text revels in sexual detail. From its first edition, the *Masterpiece* promised readers details about sexual intercourse unusual in this genre. On the title page of the first version of the *Masterpiece,* which like many contemporary title pages served as a kind of table of contents, the work promised "A word of Advice to both Sexes in the Act of Copulation."

In Northampton, as in the earlier London plays, "Aristotle" served as a figure representing the knowledge of female bodies. That knowledge seems to have been understood as the rightful property of women, but it was available to young men by virtue of print. Timothy and Simeon Root and Moses Sheldon joked about the book with Rebeccah Strong, telling her, "You need not be scared, we know as much about ye as you, and more too."[59] Mary Downing described to Jonathan Edwards a scene at her mother's house two summers earlier. Oliver Warner and some other young men had "laughed and made sport of what was about girls." Again, the young men teased the young women present by claiming to have better knowledge of female bodies than did the women: "They seemed to boast as if they knew about girls, knew what belonged to girls as well as girls themselves." Downing inferred that their boastful knowledge came from the book in question: "They seemed by their talk to apprehend that they got what they knew about girls out of that book."[60] At this moment, the *Masterpiece* functions almost as a totem, an object symbolizing these men's sudden and powerful acquisition of knowledge not usually permitted to them in their culture.[61]

Young men and women understood the *Masterpiece* as belonging in some way to women, both literally as a material object, and figuratively as a container of knowledge. Noah Baker kept the book hidden inside the lining of his coat. When his sister Sarah found it there, she promptly gave it to their mother. John Lancton, who was among the young men who teased Mary Downing at her mother's house, referred to the book as a "granny book." "Granny" was slang for "midwife," and Jonathan Edwards categorized these as "midwives books."[62] The ways in which these men and women talked about the *Masterpiece* suggest that knowledge of reproduction was considered female property.

Between the Spoken and the Written Word

Jonathan Edwards's detailed notes of his interrogations of his young parishioners reveal how this particular copy of the *Masterpiece* circulated between spoken and written words. Timothy Root, with Noah Baker and Elkanah Root, read it aloud in front of Naomi Warner and Bathsheba, a young African American woman referred to as Major Pumeroy's servant, at David Burt's house. There was laughter, and there were attempts to catch hold of the young women and kiss them.[63] Sometimes the book was used more aggressively; Medad Lyman and Oliver Warner also claimed to know "what the girls was, what nasty creatures they was."[64] At times, the book seems to have been spoken aloud in a kind of jesting or flirtatious way among men and women. At others, the men spoke from or about the book in more hostile or misogynist ways.

I have said that the book "circulated" between spoken and written words because the traffic was curiously two-way. At times, Edwards's notes suggest that the young men who teased women about their knowledge of the female body attributed more to the *Masterpiece* than was actually contained within it. Joanna (or Hannah) Clark and Bathsheba, Pumeroy's servant, were walking down the street one day when they met Oliver Warner. He said to them: "When will the moon change girls? I believe you can tell. I believe you have circles round your eyes. I believe it runs."[65] Warner was referring to a mishmash of beliefs about menstruation. It was long thought that the menses were provoked by the full moon; as the first version of the *Masterpiece* explains, "[T]he Blood being agitated by the Moons force, the Courses of Women flow from them."[66] Women were in general more subject to the influence of the moon. As Nicholas Culpeper explains, "[T]he Moon hath great influence upon all Elementary Bodies, but more upon Women than Men, because they are of her own Sex."[67] However, the detail that menstruating women had a blue circle around their eyes is not in the *Masterpiece* or any other popular medical book I have seen. It would appear to be an item of popular or local knowledge rather than something Warner learned from what he called "the young folks bible."[68] However, given the ways in which Warner and other young men claimed that their knowledge of the secrets of nature came from a granny book, it is almost as though they in turn reconstructed its contents to include the sum of their knowledge.

The third version of the *Masterpiece* itself makes a nod to the spoken word by including verses. At the end of most chapters, there is a verse that sums up the main points in a kind of singsong series of couplets. The chapter on barrenness, for example, concludes:

Which to both Sexes clearly doth relate
How Nature sometimes doth debilitate :
And likewise shews, how those who love to pry
Into the Cause of Things may soon espy
On which Side Insufficiency does lie :
And 'tis a maxim 'mong Physicians known,
The Cure's half wrought, when once the Cause is shown.
Here the Fair Sex those Remedies may see,
Which will, if barren, make them fruitful be.[69]

Verses such as these were used as memory aids. Schoolchildren, for example, chanted improving verses out loud, and chapbooks sometimes summed up their brief chapters in similar verses. There are also occasional pithy couplets within the *Masterpiece* that serve the same function. The chapter on the parts of generation in men and women notes, "Man and his Wife are but one right / Canonical Hermaphrodite," stressing the similarities between the two sexes.[70]

The Aristotle texts circulated between spoken and written words in one additional important way. In the section of the *Last Legacy* devoted to physiognomy, there are many woodcuts that illustrate various facial features.[71] One is a picture of a woman's face that illustrated the *Legacy*'s discussion of the meaning of moles on the face. However, this woodblock was already in circulation as an illustration for ballads. Those spots of her face were not originally moles but beauty patches, denoting this as a woman of fashion. Ballads such as "Love Al-a-Mode, or, The Modish-Mistris," "The Fair Maid of Islington," and "Mars and Venus" featured her image.[72] The topic of physiognomy was incorporated into the third version of the *Masterpiece*, although these illustrations were not.

The *Masterpiece*'s section on physiognomy alludes to the spoken word in yet another way. It has more verses than any other part of the book. For example, the reader is advised that the proportions of a man's nostrils predict the size of his genitals, summed up in the verse "Thus those who chiefly mind the brutal Part, / May learn to chuse a Husband by this Art."[73] Verses such as these make it easy to picture the lads of Northampton reading bits of the book out loud and teasing their female companions. Physiognomy was also the subject of ballads that proffered advice on selecting a marriage partner, although they were not usually as explicit about sexual anatomy. The ballad "The Young Mans Approbation Against the Wise Fortune-teller" described its message:

Wherein he shows to all Batchellors rare

To chuse a Wife that's civil by her hair,

Take not a red, nor a sandy do not chuse

But flaxen or brown thy love will not abuse.[74]

The third version of the *Masterpiece* would echo this ill opinion of blondes, albeit in a somewhat different key: "[I]f the Hair be white, or yellowish Colour, he is by Nature proud and bold, dull of Apprehension, soon angry, a Lover of Venery, and given to Lying, malicious, and ready to do any mischief."[75] To sum up, both the illustrations to the *Legacy*'s section on physiognomy and the *Mas-*

terpiece's use of verses in the same section link their discussions to ballads, which themselves circulated between written and sung or spoken versions.

Jonathan Edwards handled the so-called bad book controversy poorly. He preached a sermon in which he named the offenders. Tactlessly, he included some of the offspring of leading parishioners along with young men from less reputable families, thus alienating some key supporters. His own anxiety about the matter appears from his notes to be both about the sexual nature of the book itself and about the subsequent behavior of the young men involved. When members of the congregation met, the young men summoned to testify indicated their scorn for the proceeding by drinking at the tavern, playing leapfrog outside the church, and expressing themselves in colorfully direct ways. Edwards was much offended by Timothy Root's disdain and asked a number of witnesses to confirm that Root had said "I won't worship a wig" and "I don't care a turd, or I don't care a fart, for any of them."[76] Edwards did succeed in getting the church committee to obtain public confessions from Timothy and Simeon Root and from Oliver Warner. The list of names he read from the pulpit, however, had included ten young men, so Edwards cannot be said to have won the day by publicly shaming only three of them.

Aristotle's Masterpiece was a singularly popular medical book. Almost no other was in constant use from its first appearance in the seventeenth century well into the twentieth. However, it also shows us how popular medical books functioned in a variety of ways not always envisioned by their creators. These uses suggest that cheap medical books aimed at broad audiences were important elements in what we might call vernacular body culture. The ways in which people talked about and imagined their own and others' bodies were assembled from a rich array of sources that historians will never be able to recover completely. Popular medical books and the uses to which they were put, however, offer us one way to begin to understand that culture.

This analysis of the *Masterpiece* emphasizes the ways in which the work circulated both as a material object and as a representation of a kind of knowledge. Copies of the book itself were very mobile. The one in Northampton, for example, could be found on various occasions in the lining of Noah Baker's coat, at Samuel and David Burt's house or houses, in Rebeccah Strong's shop, and at Mary Downing's mother's house. The *Masterpiece* offered a paradoxical kind of knowledge. Repeatedly designated as a granny book, and explicitly directed to women, it nevertheless was read often by men. This book was almost the literal instantiation of midwifery writers' fears. It actually did encourage men to "jest and bourd" at women, in Eucharius Roeslin's words. Although the book was read by men, the knowledge contained in it was understood to be women's knowledge. While Edwards handled the controversy in Northampton poorly, it seems that many parishioners agreed with him that the young men's behavior was inappropriate, a kind of trespassing where they did not belong.

More generally, this chapter suggests that the *Masterpiece* and other Aristotle texts had very long careers in everyday parlance. From the early-seventeenth-century plays in which Aristotle is understood to be a sex expert, to the *Tatler*'s joke about Jenny Bickerstaffe reading Aristotle at school, these books served as emblems of female sexual knowledge. It is as though the figure of Aristotle, or the books that used his name, were a kind of reification of the "secrets" of women, meaning both their bodies and knowledge of those bodies. An understanding of this long history—of the ways in which the name "Aristotle" was associated with nudges and winks rather than philosophy—suggests how important printing was to body cultures. All of these associations and the jokes they engendered depended upon audiences who were familiar, not with the learned Aristotle of universities, but with jest books, plays, ballads, and the like. It was only in the world of cheap print, where beauty spots were transformed into moles and midwifery manuals sprouted physiognomy chapters, that an element of learned culture could take on such radically new meanings.

Historians have sometimes contrasted women's and men's knowledge in the early modern period, emphasizing the oral transmission of women's body cultures. Midwives were trained not from books but by apprenticeship, and many married women had attended enough births to deputize for a midwife if necessary. Men had no place in the birthing room, nor were they supposed to be knowledgeable about pregnancy and childbirth. While these basic outlines are undoubtedly correct, cheap print complicated these dichotomies. From the first midwifery manual published in English in the mid–sixteenth century, writers worried that the wrong sort of man would read their books. The ideal male reader was a husband, reading aloud to his (presumably unlettered) wife. But writers imagined the John Cannons and Oliver Warners long before these individuals were born. The *Masterpiece* seems to have been particularly appealing to male readers, to judge by the frequency with which they appear in the historical record. I have suggested that this appeal was not coincidental; the compilers of the work promised titillation as well as education by mentioning sex on the title page and including a frontispiece of a naked woman.

If the *Masterpiece* crossed boundaries of male and female knowledge, it also complicated another early modern dichotomy, that of the spoken and written word. Unlike other midwifery manuals, the Aristotle texts use a range of devices to make the printed word familiar and accessible to neophyte readers. Like chapbooks, ballads, and jest books, the Aristotle books drew upon conventions of speaking to make reading easier. The *Problems* is in question and answer format, ultimately deriving from spoken scholastic dialogues. The first version of the *Legacy* is full of snippets that invite oral performance—jokes, fortunetelling, and charms. It also uses pictures that readers might already have known from broadside ballads. To summarize its main points, the *Masterpiece* uses verses—a familiar mnemonic device often chanted aloud. The rare historical moments when we see readers actually using the *Masterpiece* suggest that this blurring of written and spoken words made the text especially inviting to read aloud.

It is too easy to suggest that these sets—male and female knowledge, spoken and written words—are actually the same. While it is true that men were more likely than women to be able to read in early modern England or colonial New England, I would not want to collapse these categories into each other. Instead, I wish to emphasize the ways in which the *Masterpiece* made these dichotomies more complicated and less absolute. The ways in which "Aristotle" got his reputation as a sex expert, and the ways in which the *Masterpiece* was read aloud and used to harass young women, suggest a model of circulation. Like broadside ballads, the Aristotle texts moved between spoken and written words. Similarly, these texts moved between all-male groups, such as some of the young men in Northampton, and all-female groups, such as Jenny Bickerstaffe's schoolmates or the intended readership of childbearing women, as well as mixed groups. While we can never fully understand the many reasons for the *Masterpiece*'s stunning popularity and longevity, some part of its success was due to the way it continued to circulate among different worlds.

NOTES

1. On Buchan, see Charles E. Rosenberg, "Medical Text and Social Context: Explaining William Buchan's *Domestic Medicine*," *Bulletin of the History of Medicine* 57 (1983): 22–42; C. J. Lawrence, "William Buchan: Medicine Laid Open," *Medical History* 19 (1975): 20–35. On eighteenth-century popular medical books more generally, see Roy Porter, ed., *The Popularization of Medicine, 1650–1850* (London: Routledge, 1992); Ludmilla Jordanova, "The Popularization of Medicine: Tissot on Onanism," *Textual Practice*

1 (1987): 68–79; Ginnie Smith, "Prescribing the Rules of Health: Self-Help and Advice in the Late Eighteenth Century," in *Patients and Practitioners,* ed. Roy Porter (Cambridge: Cambridge University Press, 1985), 249–82.

2. Previous studies of this work include Sir D'Arcy Power, "Aristotle's Masterpiece," in *The Foundations of Medical History* (Baltimore: Williams and Wilkins, 1931), 147–78; Otho Beall Jr., "*Aristotle's Masterpiece* in America: A Landmark in the Folklore of Medicine," *William and Mary Quarterly,* 3d ser., 20 (1963): 207–22; Vern L. Bullough, "An Early American Sex Manual, Or, Aristotle Who?" *Early American Literature* 7 (1973): 236–46; Janet Blackman, "Popular Theories of Generation: The Evolution of *Aristotle's Works,*" in *Health Care and Popular Medicine in Nineteenth-Century England: Essays in the Social History of Medicine,* ed. John Woodward and David Roberts (London: Croom Helm, 1977), 56–88; Angus McLaren, "The Pleasures of Procreation," in *William Hunter and the Eighteenth-Century Medical World,* ed. William F. Bynum and Roy Porter (Cambridge: Cambridge University Press, 1985), 323–420; Roy Porter, "'The Secrets of Generation Display'd': *Aristotle's Master-piece* in Eighteenth-Century England," in *'Tis Nature's Fault: Unauthorized Sexuality during the Enlightenment,* ed. Robert P. McCubbin (Cambridge: Cambridge University Press, 1985), 1–21.

3. Francis Place, *The Autobiography of Francis Place,* ed. Mary Thale (Cambridge: Cambridge University Press, 1972), 45.

4. I refer, of course, to Carlo Ginzburg's analysis of the reading practices of the Friulian miller Menocchio. Carlo Ginzburg, *The Cheese and the Worms,* trans. John and Anne Tedeschi (Baltimore: Johns Hopkins University Press, 1982). More generally, Michel de Certeau has engagingly suggested that all reading is poaching, and Roger Chartier has pioneered an understanding of the ways in which texts and practices are appropriated and reconfigured by their consumers. Michel de Certeau, *The Practice of Everyday Life* (Berkeley: University of California Press, 1984); Roger Chartier, "Culture as Appropriation: Popular Cultural Uses in Early Modern France," in *Understanding Popular Culture: Europe from the Middle Ages to the Nineteenth Century,* ed. Steven L. Kaplan (Berlin: Mouton Publishers, 1984), 230–53; idem, "Texts, Printings, Readings," in *The New Cultural History,* ed. Lynn Hunt (Berkeley: University of California Press, 1989), 154–75.

5. This estimate is based upon a larger study of popular medical works, 1640–1800, for which I have compiled a bibliography of approximately 2,500 editions from the online ESTC (English Short Title Catalog), the holdings of a number of libraries, contemporary book advertisements, and the like. I define "popular" as those works that specifically claim to be for a lay or nonmedical audience. The Aristotle texts continued to be published in large numbers in the nineteenth century, and into the twentieth.

6. James Joyce, *Ulysses,* ed. Jeri Johnson (Oxford: Oxford University Press, 1993), 226, 722.

7. A. L. Rowse, *A Cornishman at Oxford* (London: Jonathan Cape, 1965), 196.

8. Power, "Aristotle's Masterpiece," 147.

9. *The Problemes of Aristotle, with other Philosophers and Phisitions. Wherein are contayned diuers questions, with their answers, touching the estate of mans bodie* (At Edenborough: Printed by Robert Waldgraue, 1595), STC (2d ed.), 763. There is also one extant copy in private hands of a London edition of the same year. Ann Blair has shown that European books published under this title actually consist of two different works. The first is that which is still included in modern editions of Aristotle's works, although it is not considered to be Aristotle's own work. It was first published in Latin in 1475. This text

was never to my knowledge published in English, and hence it does not form part of this analysis. The other *Problems* was first published in Latin in 1483 or 1488, rapidly translated into various European vernaculars, and frequently republished in the sixteenth and seventeenth centuries. I am very grateful to Professor Blair for sharing her work on these texts with me; see her "Authorship in the Popular 'Problemata Aristotelis,'" *Early Science and Medicine* 4 (1999): 189–227.

10. *Masterpiece 1: Aristotle's Master-piece: or, The Secrets of Generation Displayed* (London: printed for J. How, and are to be sold next door to the Anchor Tavern in Sweethings Rents in Cornhil, 1684), Wing (2d ed., 1994), A3697fA; *Masterpiece 2: Aristotle's Master-piece Compleated* (London: printed by B.H., and are to be sold by most booksellers, 1697), Wing (CD-ROM, 1996), A3697kA; *Masterpiece 3* is first published under a different name: *Insigne artificium Aristotelis: or, Aristotle's Compleat Master-piece. In Two Parts. Displaying the Secrets of Nature in the Generation of Man. . . . To which is added, Hippocrates his treasure of health: or, family physician* (London: printed, and are to be sold by the booksellers, 1702). I know of only one extant copy, at the British Library, shelfmark RB.23.a.7258. Although no publisher is listed, the work includes an advertisement for Eben Tracy, a bookseller on London Bridge. The earliest edition I have found under the usual name is *Aristotle's Compleat Master-piece. In Three Parts. Displaying the Secrets of Nature in the Generation of Man . . . to Which Is Added, a Treasure of Health; Or, the Family Physician*, 12th ed. (London: printed, and sold by the booksellers, n.d.). It appears to date from the 1710s or 1720s.

11. *Legacy 1: Aristotle's Legacy: Or, His Golden Cabinet of Secrets Opened. In five treatises. 1. the wheel of fortune. 2. the art of palmestry. 3. a treatise of moles. 4. the interpreter of dreams. 5. observations on fortunate and unfortunate days. with many other secrets and experiments never before published . . .* (London: printed for J. Blare, at the looking glass on London Bridge, ca. 1690?); *Legacy 2: Aristotle's Last Legacy, Unfolding the Mystery of Nature in the Generation of Man: treating, I. Of virginity, . . . IX. Excellent remedies against all diseases incident to virgins and child-bearing women* (London: printed for A. Bettesworth and C. Hitch, and J. Osborne, and T. Hodges, [1730?]).

12. *Aristotle's Compleat and Experienc'd Midwife* (London: printed, and sold by the booksellers, 1700), Wing (CD-ROM, 1996), A3697bA.

13. *Aristotle's New Book of Problems, Set Forth by Way of Question and Answer. To which are added, a great number from other famous philosophers, astrologers, astronomers, and physicians. Shewing the secrets of nature and art: together with the interpretation of dreams*, 6th ed. (London: printed for John Marshall, 1725).

14. *Aristotle's Manual of Choice Secrets, Shewing the Whole Mystery of Generation. With receipts to prevent barrenness, and cause conception. Very necessary to be known and practiced by all midwives, nurses, & young married women. Translated out of Latin by J.P.* (London: printed for John Back, at the Black-Boy on London-Bridge, 1699), Wing (CD-ROM, 1996), A3697eA.

15. W.S., *A Family Jewel, or the Womans Councellor: Containing, I. An exact method of preventing or curing all diseases, and grievances incident to children, . . . V. The art of japanning and painting in oil* (London: printed and sold by A. Baldwin, 1704 [1705]). The National Library of Medicine copy lacks any version of the section V referred to in the title.

16. Levinus Lemnius, *The Secret Miracles of Nature* (London: printed by Jo. Streater, and are to be sold by Humphrey Moseley at the Prince's Arms in S. Paul's Church-Yard, John Sweeting at the Angel in Popes-Head-Alley, John Clark at Mercers-Chappel, and

George Sawbridge at the Bible on Ludgate-Hill, 1658), Wing L1044; T.C., I.D., M.S., T.B., *The Compleat Midwifes Practice* (London: Printed for Nathaniel Brooke at the Angell in Cornhill, 1656), Thomason Tracts E.1588[3].

17. See William Eamon, *Science and the Secrets of Nature* (Princeton: Princeton University Press, 1994), 275–76, for a brief discussion of this work.

18. Levinus Lemnius, *A Discourse Touching Generation. Collected out of Lævinus Lemnius, a most learned Physician* (London: printed by John Streater, 1664). Identified as Wing L1043A on UMI microfilm.

19. John Sadler, *The Sicke VVoman's Private Looking-Glasse* (London: Printed by Anne Griffin, for Philemon Stephens, and Christopher Meridith, 1636), STC (2d ed.), 21544.

20. Nicholas Culpeper, *A Directory for Midwives* (London: Printed by Peter Cole, at the sign of the Printing-Press in Cornhill, near the Royal Exchange, 1651), Thomason Tracts E.1340[1]; John Pechey, *The Compleat Midwife's Practice Enlarged . . . The fifth edition corrected, and much enlarged, by J.P. Fellow of the College of Physicians, London* (London: printed for R. Bentley in Russel-street, Covent-Garden, H. Rhodes at the corner of Bride-Lane, in Fleet-street, J. Philips, at the King's Arms, and J. Taylor at the Ship in St. Paul's Church-Yard, 1697), not in Wing.

21. I say "appear to be" because I have not found another source for these sections. Although they are new to the *Masterpiece*, I suspect that they derive from another source.

22. Paula McDowell, *The Women of Grub Street: Press, Politics, and Gender in the London Literary Marketplace, 1678–1730* (Oxford: Clarendon Press, 1998); Adam Fox, *Oral and Literate Culture in England, 1500–1700* (Oxford: Clarendon Press, 2000).

23. [Thomas Middleton], *The Famelie of Loue. Acted by the children of his Maiesties Reuells* (At London: printed [by Richard Bradock] for Iohn Helmes, and are to be sold in Saint Dunstans Churchyard in Fleetstreet, 1608), STC (2d ed.), 17879, sig. F3r, act 4, scene 2.

24. For example, Nicholas Culpeper lists a total of fifteen different signs of pregnancy, ranging from changes in the veins in the breast to diagnostic tests involving the woman's urine. Culpeper, *Directory for Midwives*, 126–28.

25. J. Oldham, "On Pleading the Belly: A History of the Jury of Matrons," *Criminal Justice History* 6 (1985): 1–64.

26. Laura Gowing, "Secret Births and Infanticide in Seventeenth-Century England," *Past and Present*, no. 156 (1997): 87–115; P. C. Hoffer and N. E. H. Hull, *Murdering Mothers: Infanticide in England and New England, 1558–1903* (New York: New York University Press, 1981).

27. Thomas Dekker, *The Honest Whore, With, the Humours of the Patient Man, and the Longing VVife* (London : Printed by V[alentine] S[immes and others] for Iohn Hodgets, and are to be solde at his shop in Paules church-yard, 1604), sig. B1r, act 1, scene 1.

28. On Albertus Magnus, see Helen Rodnite Lemay, trans., *Women's Secrets: A Translation of Pseudo–Albertus Magnus's "De Secretis Mulierum" with Commentaries* (Albany: State University of New York Press, 1992). Albertus resembles the Aristotle texts in a number of ways: it too was published along with other works of Albertus, such as the one on stones and minerals, and, of course, the title implies authorship by an authority who, if not ancient, carried similar connotations of antique learning. However, Albertus was not published often in English. There were a few seventeenth-century editions, and Edmund Curll, printer of scurrilous and pornographic texts, brought out an edition in 1725.

29. Edward Ravenscroft, *The London Cockolds. A comedy; as it is acted at The Duke's Theatre* (London: printed for Jos. Hindmarsh at the Sign of the Black-Bull near the Royal-Exchange in Cornhill, 1682), Wing (2d ed., 1994), R332, p. 2, act 1, scene 1.

30. On the relationships between spoken and written words in the Middle Ages, see Michael Clanchy, *From Memory to Written Record: England, 1066–1307* (London: Edward Arnold, 1979).

31. *The Problemes of Aristotle, with other Philosophers and Phisitions* (At London: Printed by Arnold Hatfield, 1597), STC (2d ed.), 764, sig. e5r.

32. *The Tatler*, ed. Donald F. Bond (Oxford: Clarendon Press, 1987), 1:469. This letter, no. 68, is dated September 1709. Many thanks to Ann Kelly for this reference.

33. Ibid., 1:512–13 (letter no. 75).

34. *Onania; or the Heinous Sin of Self-Pollution* (London: Printed, Re-printed at Boston; for John Phillips, and Sold at his Shop on the South Side of the Town House, 1724), 163. This is the edition most easily available in reprint: *The Secret Vice Exposed! Some Arguments Against Masturbation* (New York: Arno Press, 1974).

35. On gossips, see David Cressy, *Birth, Marriage, and Death: Ritual, Religion, and the Life-Cycle in Tudor and Stuart England* (Oxford: Oxford University Press, 1997), 55–59, 84–87.

36. Tim Hitchcock, "Sociability and Misogyny in the Life of John Cannon, 1684–1743," in *English Masculinities, 1660–1800*, ed. Tim Hitchcock and Michele Cohen (London: Longman, 1999), 25–43.

37. John Cannon, "Memoirs of the Birth, Education, Life and Death of: Mr. John Cannon. Sometime Excise Officer & Writing Master at Mere Glastenbury & West Lydford in the County of Somerset," 1743, Somerset Record Office, MS DD/SAS C/1193/4, p. 29.

38. Ibid., 41.

39. Ibid., 145. See a fuller discussion of this incident in Hitchcock, "Sociability and Misogyny," 38.

40. For further discussion of this image, see my essay "Hairy Women, Naked Truths, and Sexuality in Cheap Print," forthcoming in the *William and Mary Quarterly*.

41. On these images, see Tamsyn Williams, "'Magnetic Figures': Polemical Prints of the English Revolution," in *Renaissance Bodies: The Human Figure in English Culture, c. 1540–1660*, ed. Lucy Gent and Nigel Llewellyn (London: Reaktion Books, 1990), 86–110.

42. Scholars have debated whether books such as the *Masterpiece* should be categorized as pornographic. See, for example, Roy Porter, "Spreading Carnal Knowledge or Selling Dirt Cheap? Nicolas Venette's *Tableau de l'amour conjugal* in Eighteenth-Century England," *Journal of European Studies* 14 (1984): 233–56. Roger Thompson elides medical books with a range of other bawdy or pornographic texts, without much consideration of the readerships of different genres or their places in the print marketplace: *Unfit for Modest Ears: A Study of Pornographic, Obscene, and Bawdy Works Written or Published in England in the Second Half of the Seventeenth Century* (Totowa, N.J.: Rowman and Littlefield, 1979). I prefer not to use the category "pornographic" in relation to this text because it seems that pornography as such is a genre characterized by a much more elaborate (and less medical) frame and was produced to be consumed by a small and wealthy audience. See Lynn Hunt, ed., *The Invention of Pornography: Obscenity and the Origins of Modernity* (New York: Zone Books, 1993); Tim Hitchcock, *English Sexualities, 1700–1800* (London: Macmillan, 1997), 12–23; David Foxon, *Libertine Literature in England, 1660–1745* (New Hyde Park, N.Y.: University Books, 1966).

43. The quotation is from another parishioner, Samuel Hopkins: Samuel Hopkins, *The Life and Character of the Late Reverend Mr. Jonathan Edwards* (Boston: Printed and sold by S. Kneeland, opposite to the Probate-Office in Queen-Street, 1765), 53. I am very grateful to Ava Chamberlain for sharing much information about the so-called bad book controversy with me. Ava Chamberlain, "Bad Books and Bad Boys: The Transformation of Gender in Eighteenth-Century Northampton," *New England Quarterly* (forthcoming). Some of the evidence has been published in John E. Smith, Henry S. Stout, and Kenneth P. Minkema, eds., *A Jonathan Edwards Reader* (New Haven: Yale University Press, 1995), 172–78, as well as in Thomas H. Johnson, "Jonathan Edwards and the 'Young Folks Bible,'" *New England Quarterly* 5 (1932): 37–54. The MSS sources for the controversy are in the Franklin Trask Library, Andover-Newton Theological School, Newton Center, Mass., folder ND1, items 10B and 11. Transcripts on deposit at office of *The Works of Jonathan Edwards*, Yale University. I am deeply grateful to the Franklin Trask Library and to the editors of the Edwards papers for permission to cite these documents, and to Ava Chamberlain for sharing her knowledge of them with me. On Edwards's understanding of the processes of reproduction, see Ava Chamberlain, "The Immaculate Ovum: Jonathan Edwards and the Construction of the Female Body," *William and Mary Quarterly*, 3d ser., 57 (2000): 289–322.

44. In addition those cited above, see Sereno Edwards Dwight, *The Life of President Edwards* (New York: G. & C. & H. Carvill, 1830), 299–300; Patricia J. Tracy, *Jonathan Edwards, Pastor: Religion and Society in Eighteenth-Century Northampton* (New York: Hill and Wang, 1979), 160–64; Ola Elizabeth Winslow, *Jonathan Edwards, 1703–1758* (1940; reprint, New York: Collier Books, 1961), 203–11.

45. Four editions of the *Masterpiece* were published by B.H. between 1697 and 1707. I cannot be certain that Harris was B.H. However, he is the only known London publisher with the initials B.H., and he had strong personal links to a number of other Aristotle publishers. Both John How and Thomas Norris, who produced various Aristotle texts, were apprenticed to him.

46. James N. Green, "'The Cowl knows best what will suit in Virginia': Parson Weems on Southern Readers," *Printing History* 17 (1995): 26–34, quotation on 34. Many thanks to Jim Green for sharing this reference with me.

47. Edwards transcript, 9, 12.

48. Thomas Dawkes, *The Midwife Rightly Instructed: Or, the Way, Which All Women Desirous to Learn, Should Take, to Acquire the True Knowledge and Be Successful in . . . the Art of Midwifery* (London: printed for J. Oswald, at the Rose and Crown in the Poultry, near Stocks-market, 1736).

49. Edwards transcript, 9.

50. Culpeper's *Directory* is something of an exception to this rule. Although it is dedicated to the midwives of England, it deploys visual metaphors for knowledge, emphasizing anatomical truths verified by witnessing, rather than touching.

51. John Maubray, *The Female Physician, Containing All the Diseases Incident to That Sex, in Virgins, Wives, and Widows; Together with Their Causes and Symptoms, Their Degrees of Danger, and Respective Methods of Prevention and Cure* (London: printed for James Holland, at the Bible and Ball in St. Paul's Churchyard, 1724). A second edition was published in 1730.

52. On the use of instruments in midwifery, see Adrian Wilson, *The Making of Man-Midwifery: Childbirth in England, 1660–1770* (Cambridge: Harvard University Press,

1995). The fillet is a sort of loop of flexible material attached to a rigid handle. It is roughly analogous to forceps in function; both could be placed around a baby's head to provide traction to promote delivery, although practitioners often argued for one of the two instruments. Deventer advocated for the use of a fillet-like device, but only, it seems, for the delivery of a dead child. In general, he promoted manual rather than instrumental intervention.

53. Dawkes, *Midwife Rightly Instructed*, xvii–xviii.

54. Eucharius Roeslin, *The Birth of Mankind, otherwise called The Womans Book*, 4th ed. (London: Printed for J. L. Henry Hood, Abel Roper, and Richard Tomlins, and are to be sold at their Shops in Fleetstreet; and at the Sun and Bible in Pie-Corner, 1654), Wing (2d ed., 1994), R1782B, 8. Roeslin was first published in German in 1513 and in English in 1540. Thomas Raynalde, one of the work's English translators, is often listed as the author. "Bourd" is an archaic usage; it means to accost or make advances towards.

55. Ibid., 12.

56. Jacques Guillemeau, *Child-birth, or the Happy Deliverie of Women* (London: Printed by A. Hatfield, 1612), STC (2d ed.), 12496, sig. 2v. Guillemeau was first published in Paris in 1609.

57. Ibid., sig. 3r. The translator also defends himself by pointing out that Guillemeau himself wrote in the vernacular.

58. *Masterpiece 3*, sig. A4r. This paragraph is taken directly from one of the sources of the *Masterpiece*, namely the *Discourse Touching Generation*.

59. Edwards transcript, 9.

60. Ibid., 12.

61. David Cressy, "Books as Totems in Seventeenth Century England and New England," *Journal of Library History* 21 (1986): 92–106. On the cultures of books and reading in colonial New England, see Hugh Amory and David D. Hall, eds., *The Colonial Book in the Atlantic World* ([Worcester, Mass.]: American Antiquarian Society, 2000); David D. Hall, *Cultures of Print: Essays in the History of the Book* (Amherst: University of Massachusetts Press, 1996).

62. Edwards transcript, 9, 13, 14.

63. Ibid., 10.

64. Ibid., 11.

65. Ibid., 10; see also 18.

66. *Masterpiece 1*, 69. Jane Sharp, however, refers to another common belief, that young women menstruated at the new moon but older women at the full moon. Jane Sharp, *The Midwives Book* (London: printed for Simon Miller at the Star at the West End of St. Pauls, 1671), Wing (2d ed., 1994), S2969B, 76. More generally, see Patricia Crawford, "Attitudes to Menstruation in Seventeenth-Century England," *Past and Present*, no. 91 (1981): 47–73.

67. Culpeper, *Directory for Midwives*, 96.

68. Edwards transcript, 9.

69. *Masterpiece 3*, 58.

70. *Masterpiece 3*, 15.

71. *Aristotle's Legacy: Or, His Golden Cabinet of Secrets Opened. In Five Treatices* ([London]: Printed for J. Blare, at the Looking Glass on London Bridge, [1699]), Wing (CD-ROM, 1996), A3697dA.

72. "The Fair Maid of Islington; or The London Vintner Over-Reach'd" ([London]:

printed for J. Clarke, W. Thackeray, and T. Passenger, [between 1684 and 1686]), Wing (CD-ROM, 1996), F100B; "Love Al-a-Mode, or, The Modish-Mistris" (n.p.: ca. 1685), Wing (CD-ROM, 1996), L3193A; "Mars and Venus: Or, the Amorous Combatants," ([London]: Printed for J. Wright, J. Clarke, and T. Passenger, [between 1672 and 1685]), Wing (CD-ROM, 1996), M721.

73. *Masterpiece 3*, 106.

74. *The Young Mans Approbation Against the Wise Fortune-teller* ([London]: Printed by J. Lock for J. Clarke at the Bible and Harp in West-Smith-Feild, c. 1674?), Wing (CD-ROM, 1996), W176; *The Famous History of Fryer Bacon* (London: printed by E. Cotes, for F. Grove dwelling upon Snow-hil, 1661), Wing (2d ed., 1994), F371, sig. C1v–C2r.

75. *Masterpiece 3*, 100. The text explains that the advice about men's appearance is applicable to women as well.

76. Edwards transcript, 21.

4

KATHLEEN BROWN

The Maternal Physician
Teaching American Mothers to Put the Baby in the Bathwater

In 1811, New York publisher Isaac Riley released a pocket-sized volume of medical advice dedicated to the care of infants and young children. *The Maternal Physician: A Treatise on the Nurture and Management of Infants, from the Birth until Two Years Old* claims the dual distinction of being the first child care manual written and published in the United States and one of the earliest American health guides of any kind. It also appears to have been the first book of medical advice penned by an American woman, who identified herself on the title page as only "an American matron." Offering extracts from "the most approved medical authors" as well as the fruits of her "sixteen years' experience in the nursery," the anonymous author hoped to provide readers "with a concise and simple statement of the characteristic symptoms of the various complaints to which children are subject, in so small a compass that it may become a pocket companion." Although much about the book's publishing history remains unknown, we do know that its New York edition sold well enough to justify the publication of a second edition in Philadelphia in 1818. Nineteen years later, when Lydia Maria Child completed her volume of popular medical advice, *The Family Nurse, The Maternal Physician* was still relevant enough to be part of the corpus of works quoted, although Child remembered its anonymous author as a male doctor rather than as an American matron. Only recently have scholars identified its author as Mary Hunt Palmer Tyler, wife of

Royall Tyler, whose popular play *The Contrast* urged Americans to redefine ideals of manhood.[1]

I stumbled on *The Maternal Physician* while researching a project on the history of cleanliness in early America. Initially, I was interested in what it had to say about the importance of washing infants daily in water. After reading dozens of eighteenth- and early-nineteenth-century popular medical advice books by British and American doctors—every one male—and dozens of household advice books, cookbooks, and children's books, almost all written by women, I increasingly began to view Tyler's book as an outlier: a popular medical guide written by a woman in the early nineteenth century, focused not on midwifery, as was the case with many female-authored advice books from the seventeenth century, but on the care of infants and children, a genre that male doctors had previously monopolized.[2]

Interpreted as an artifact of print culture, *The Maternal Physician* offers a glimpse of the challenges of popular medical writing, female authorship, and publishing during the first decades of a new nation self-conscious of its need to claim cultural independence in the aftermath of its political independence. In trying to make sense of this book in this context, I asked the following: What were the touchstones of popular medical belief and practice that Tyler used to establish credibility as the author of a medical text? What did she claim were the unique features of her advice, the reasons why someone should buy and read her book? How did her femaleness figure in her claims to credibility and medical authority? *The Maternal Physician* is also a document in social history, revealing how one woman valued her life experience as a healer and tried to reconcile it with the book knowledge she had accumulated as a reader of medical texts. With her primary duties defined by the affective bonds of patriotism and motherhood, Tyler navigated the competing claims to legitimacy of medical knowledge derived from texts and that contained in folk wisdom, of male academic medical training and female traditions of healing, at a time when clear winners had yet to emerge. Tyler's advice book provides some sense of what was at stake for mothers like herself and the ways they turned selectively to print culture to protect their children's health.

In addition to analyzing Tyler's book in these two interconnected ways—as an artifact of print culture and as a social history source on one woman's child care practices, I hope to use it to raise more general questions about changing regimes for caring for the body and the relationship between maternal and medical authority in the young nation. What can *The Maternal Physician*, a

book of only modest commercial success, tell us about the relationship between maternal and infant bodies in early American society? What can it tell
us about women's authority as healers in a world in which male doctors were
beginning to challenge the claims of midwives to deliver babies?

I want to turn now to some of the contexts useful to understanding Tyler's
work: print culture, especially the genre of popular health manuals and the textual strategies used by authors to establish credibility; female traditions in the
healing arts embodied in the receipt or recipe book; and the legacy of the
American Revolution for health and cleanliness in the new nation.

Print Culture

The genre of popular health manuals is arguably the most important context for understanding the choices Mary Tyler made about her book's form and
content and how she positioned herself in relation to the authors of other medical guides. In the latter half of the eighteenth century, British popular health
publications directed at women were usually written in response to high infant
mortality rates in London. Medical writers such as Hugh Smith and Michael
Underwood interpreted infant death as a loss of valuable resources needed by
a nation or empire. Citing the same infant mortality statistics, both men noted
that two-thirds of the children born in London were "lost to society" soon after
birth, and nearly three-quarters by the age of two. They recommended maternal breastfeeding and keeping the infant's body, clothing, and environment
clean as remedies. Increasingly, these authors aimed their texts at female audiences and used a variety of techniques to make them more accessible to lay
readers. They included poetry in the books, removed technical explanations
and Latin terms, employed an epistolary style, focused on using household resources for health maintenance, and instructed with catechisms. They implicitly—and sometimes explicitly—connected maternal responsibility for infant
health with national and imperial power, a message that would take a somewhat different form in Tyler's American guide.[3]

Surveying this popular health literature reminds us of Tyler's anomalous position as a female author. Even male authors in this genre were at pains to
claim legitimacy, owing to the lack of a single accepted source of medical authority—a context that potentially opened up opportunities for women to present rival claims. Having trespassed onto intellectual terrain already staked out

by credentialed men, however, female authors appear to have borne an even greater burden. Yet as we will see, Tyler employed strategies common to authors of both sexes. Both male and female authors offered formulaic apologies for their work, to cite one common example. In the preface, an author might beg the reader's forgiveness for the intrusion of his or her opinion, offering a deferential counterbalance to the more assertive tone of the main text. Authors of both sexes also attempted to deny the market for books as a motive for publishing, explaining that friends had urged them to make such helpful advice available to the public. This unselfish motive stood in implicit contrast to others like self-promotion or moneymaking. Deference and commercial self-effacement made the advice giver appear a disinterested and therefore more trustworthy provider of medical advice. In addition to using these time-honored strategies, authors of both sexes claimed to offer advice tailor-made to a certain national situation. For American authors in the new nation, such claims performed the linked and vital jobs of building readership and strengthening the nation.[4]

Female authors elaborated upon these rhetorical strategies by combining them with explicit claims about their qualifications as women who had profited from life experience. This strategy worked best for authors who claimed expertise in areas most readers already believed belonged to women—cooking, conduct, virtue and morality, and the perils of courtship—and less well for topics further removed from these explicitly female subjects—for example, histories or medical advice books. In these latter cases, authors adumbrated the ways womanhood gave them unique insights.[5]

Tyler was hardly the first author to celebrate a special role for enlightened mothers in promoting health, even if she was one of the first English-speaking women to do so in print. Several prominent eighteenth-century political and medical writers commented on the domestic virtue of educated mothers and its importance for both the political health of the state and the bodily health of its individual citizens. Identifying virtuous motherhood as a woman's natural calling and the private family as her natural place, Jean Jacques Rousseau urged that women "deign to nurse their own children" and shun both nightlife and public activity. Medical writers like Tissot envisioned educated, enlightened rural matrons contributing to the health of the common people by dispensing advice, scientific knowledge, and medicine. In William Buchan's popular works, rational mothers dispensed with unhealthy customs and brought enlightened, healthful practices to their households.[6] But for Tyler, and perhaps also for her female readers, there was a difference between dutifully performing a role prescribed by a male doctor, which might result in a local reputation for benevolence, and disseminating medical advice oneself in print to a potentially wide readership. In doing the latter, a female author ventured a public claim to originality and to medical authority, both of which were unusual for women in 1811. As we will see, in her text Tyler vacillated between repudiating and embracing this claim. But there was no question that as author rather than consumer of a medical guidebook, Tyler was positioned not just to receive the truth but to be accountable for her articulation of it.

The Receipt Book Tradition

Placing Tyler in the context of other authors reminds us that textual authority was nearly always constructed through references to other sources of authority—sources represented in, but not wholly contained by, the printed text. Frequently, we see references to a corpus of "experience": a doctor's encounters with thousands of sick patients, the mistress of the household's preparation of thousands of meals, or the midwife's delivery of thousands of babies. To access this life experience in print culture—to use it as evidence to advance a particular point of view—an author had to sort through it, identifying patterns and resolving contradictions that might lead to competing truth claims. Tyler's anomalous position as a female author of a popular medical advice book—one of the few such authors not trained as a doctor—raises questions about what other models were available to her for defining her life experience as a useful source of medical advice.

One obvious model Tyler was likely familiar with was the receipt book or recipe book. Receipt books embodied the domestic medical authority of the housewife and the local healer who was not academically trained. Compiling conflicting pieces of advice derived from contradictory first principles and methods, receipt book authors retained a strong connection to folk traditions for healing, even as they incorporated remedies and methods from academically trained doctors.

Claims to efficacy—what would subsequently be dismissed as rank empiricism by academically trained physicians—were the ultimate test of all remedies and had the potential to challenge the primary claim to credibility: the social position or educational privilege of a particular author. A recipe from an unschooled person low in the social order might thus present itself in such a book as equal in value to that of a wealthy landowner's or doctor's cure. This was the leveling potential of the receipt book.[7]

Well into the nineteenth century, collections of receipts remained the most common textual sources for traditions of female healing (although men occasionally kept them too) and our best documentary evidence for how they constructed their authority as healers outside the realm of print culture. One particularly rich early American example of this tradition is the receipt book of Elizabeth Coates Paschall. The daughter of a prominent Philadelphia Quaker family and wife of merchant Joseph Paschall, Paschall was born in 1702 and began keeping a receipt book sometime in her adulthood, most likely after she was married in 1721. Like other receipt books, Paschall's is an eclectic collection containing recipes for pickled oysters and potted venison, directions for making an ointment to relieve sore nipples, and a cure for "Pissing a Bed."[8]

Paschall's recipes came from a variety of sources. She included recipes given to her by itinerant strangers as well as by neighbors, carefully noting the anecdotal evidence of efficacy in each case. In one instance, she noted that a cure for a scald head recommended by John Holmes came from "a small Manuscript Book of his fathers Experiences." She recorded doctors' remedies in the same fashion as other recipes, providing a detailed account of how well they worked. In addition to testimonials and secondhand success stories, many of the recipes appear to have passed the difficult test of Paschall's own experience, which she also recounted in the recipe. In at least one case, she included a remedy for fever and ague—secretly notching a stick without the patient's knowledge—that was reputed to work through "simpathy."[9]

Paschall's book was not simply a written record of remedies transmitted through oral culture and other receipt books, however. As time passed she included a growing number of remedies gleaned from newspapers and magazines—the *Pennsylvania Gazette* and the *London Magazine*, for example—and from medical texts. From the second edition of Robert Boyle's posthumous three-volume *Philosophical Works*, which appeared in 1738, she extracted a cure for the "stone" that Boyle had himself culled from Richard Ligon's *History of Barbadoes*.[10] John Quincy's *Compleat English Dispensatory*, published in 1718, furnished several recipes for healing plasters and salves. This integration of print culture sources into a collection of receipts distinguished Paschall's compilation from many earlier examples in this genre. It also anticipated Tyler's efforts to build maternal medical authority on a combination of book learning and experience.

Paschall's eclectic healing methods appear to have both competed with and complemented those of trained doctors, whom she regarded variously as a resource useful in cases of serious illness, of limited utility in terminal cases, and one to be avoided if possible. Paschall included some recipes because she believed they would make a physician's intervention unnecessary. Other recipes appear in her book because she had occasion to recommend them to grateful doctors whose own remedies were failing. In still other cases, a remedy's efficacy became clear only after doctors had pronounced a case hopeless. Most of the doctors mentioned in Paschall's text fit this final category and appear in her pages as nameless failures who prematurely declared cases incurable. Only those with remedies of proven value—that is to say, those whose cures made it into her book—were dignified with last names.

In each instance, however, efficacy was the main test of a recipe's credibility. Doctors in particular, and socially prominent or formally educated people in general, appear to have had little inherent authority to heal in her view. All remedies, no matter what their source, were subject to her unrelenting empiricism. Indeed, if Paschall gave anyone the benefit of the doubt, it was people whose remedies she trusted for having passed the test of efficacy: Indians, country people, and elderly women.

Much as empiricism—defined here as the trial-and-error method and the "experiential" data it yielded—accorded little inherent authority to a remedy because of its source or method, it also provided little guidance for sifting through competing claims to efficacy. Paschall's book contained several different remedies for the same ailment without any indication that one method or theory of healing was more trustworthy than another. Deciding which remedy to try first in situations of serious illness would have been an analytic process of the utmost importance, but Paschall's receipt book offered few clues about this process. What the book does record is her resort to multiple remedies in difficult cases.

Paschall's system for deciding between remedies may lie in the details of patient symptoms, the administration of medicine, and the road to recovery recorded in her book. She gathered information about sick people carefully, noting details about the sequence and strength of symptoms, both those she could observe directly and those reported to her. She described precisely how a medicine was prepared and administered, and the signs that the patient was returning to health. What is not reflected in the text is how she might have sifted through these information-laden portions of the receipt, attempting to match the circumstances of a patient's illness to previous cases she had cured.

Comparing Tyler's *Maternal Physician* to the receipt book tradition exemplified by Paschall—with its negotiation of oral and print cultures, its medical eclecticism, and its unwillingness to recognize doctors as authorities except if their remedies passed the test of efficacy—helps us evaluate Tyler's connection to female healing traditions and the impact of print culture on her representation of her own authority. Writing over half a century after Paschall, Tyler was more a product of print culture than her predecessor. But her work also reflected many of the features of the receipt book tradition.

Cleanliness

Tyler was raised in a society in which cleanliness was more a question of morality, especially sexual morality, than of literal cleanliness. When colonial Americans did concern themselves with literal cleanliness, they sought it in the domestic resources—clean shirts and bed linens—created by women's labor. Personal cleanliness during this period was achieved mainly through friction and linen rather than through soap and water. Early Americans paid more attention to cleaning the garments worn by the body, where the skin's toxic effusions were believed to rub off and accumulate, than to cleaning the body itself.[11]

The American Revolution temporarily disrupted these connections between morality, domestic labor, and literal cleanliness. Linen-centered standards of cleanliness were impossible to maintain during the war, owing to short supplies of clothing and soap and soldiers' resistance to the discipline imposed by officers. Refusing to wash their own shirts, obey regulations for latrine use, or sweep out their tents, ragged, lice-infested soldiers became the object of scorn from civilians and officers alike.

After the revolution, the negative connotations of cleanliness grew as reaction against European and especially French-style luxury became more pronounced. First, there was the soldiers' own antipathy to the ways of gentlemen, as evidenced by the popularity of poems, plays, pamphlets, and songs mocking aristocratic pretensions.[12] Mary Tyler's husband, Royall, made this critique explicit in the play *The Contrast* (1787) by spoofing the imitator of British affectations with the character of Dimple and holding up the virtuous Colonel

Manly, a war veteran, as an exemplar of American male character.[13] Anti-French sentiment also appeared in artists' representations of bathing as morally suspect. In the aftermath of the French Revolution, engravers offered images of bathing that connoted vice, seduction, and vulnerability, none of which were compatible with the virtues of citizenship in the new nation. Tainted with associations of French moral corruption and the sexual license suggested by nudity, bathing suggested questionable morals and vulnerability. Depicted naked and subject to the gaze of the onlooker, children and women were the most frequent subjects of bathing scenes. While appropriate when fixed upon social dependents and inferiors, such a gaze upon a naked adult man clashed with other messages about independent male virtue. Even in depictions of women bathing, the virtue required of republican mothers and wives seemed to be at risk as the female body revealed itself to a male onlooker.[14] All these bathing images, however, contained countervailing messages about whiteness, which had long been associated with cleanliness, gentility, and purity in European portraiture, embodying it in the clean, nude bodies of women and children.

Second, there is the all-important context of republican motherhood, with its intellectual opportunities for women, and the larger reaction against the European curriculum, including Latin and Greek. Reformers like Benjamin Rush advocated abandoning Latin and Greek as remnants of aristocratic education that mystified knowledge in ways that worked to women's disadvantage. Seen as unfit to master the classics, women remained at an educational disadvantage so long as the ancient languages were a standard part of the American curriculum.[15] Authors like Tyler, who eschewed technical Latin terms in the tradition of popular medical advice books, represented the potential opportunities for women in the emphasis on practical application and efficacy rather than on theory and academic training. Educators debated about how this difference in emphasis affected a host of academic subjects, including the sciences.

When we take stock of the legacy of the American Revolution for Mary Tyler, we need to be mindful of her contemporaries' desires to remember the revolution as morally virtuous, which was difficult for late-eighteenth-century Americans to reconcile with the actual wartime conditions of bodily filth, lice, and disease. Filth and squalor were not simply erased from historical memory of the revolution, however; they were refigured in memoirs and essays after the revolution as the consequence of the brutality of British prisons or of Native American nastiness and brutishness. Both of these tropes valued cleanliness positively as a military, masculine, and moral virtue crucial for the success of the American nation. On the other hand, there was the reaction against European luxury, particularly French culture, and effeminate manhood that impeded the reception of cosmopolitan ideas about washing, bathing, grooming, and changing linens. This reaction contained a different message about republican virtue. True gentility might consist of just and polite treatment of all people, regardless of personal habits and grooming.

Mary Tyler and *The Maternal Physician*

Tyler grew up during the early years of American nationhood, with its self-conscious efforts to fashion American intellectual and cultural independence. Born in Boston in 1775 just a month before the battle of Lexington and Concord, she identified her own coming of age with that of the nation. She was educated by her mother, Elizabeth Hunt, who had received a mediocre education in the public schools but had expanded her intellect through the tutelage of her husband-to-be, Joseph Palmer. Tyler's mother and father embraced Enlightenment ideals for women's education that later found expression in the American celebration of the republican mother. This was in keeping with their militant patriotism. Joseph Palmer, Tyler's father, was a "high Whig" who dressed as an Indian to dump tea in the Boston Harbor at the Boston Tea Party. Both her father and her grandfather served as generals during the Revolutionary War. Tyler's autobiography included several letters from her father in which he commented on her writing style, urging her to avoid ornamental expressions, advice she did not heed when writing *The Maternal Physician*.[16]

Tyler eventually married a friend of her father's, the lawyer Royall Tyler. Tyler had been engaged to the daughter of Abigail and John Adams, but the match had broken up over his rakish reputation, leaving him in a severe depression. He relocated to New York City, where he wrote *The Contrast*, one of the earliest comedies written by an American and produced by a professional company of players. Royall's mother's opposition to the marriage of her son to Mary Palmer forced the couple to keep the 1794 union secret for two years. Royall subsequently wrote a novel, *The Algerine Captive*, in which, among other things, he satirized college education and medical quackery in the North and slavery in the South. During this time, the Tylers moved to Vermont, where Royall assumed responsibilities as attorney general, judge, and eventually supreme court justice.

In the year that Mary Tyler likely completed work on *The Maternal Physician*, she was thirty-five years old, living in rural Vermont, married to a supreme court justice, and the mother of eight children. In many ways she seemed to embody the ideal reader of medical guidebook authors Tissot and Buchan: an educated rural matron, dispensing medical knowledge to the uneducated peasantry. One other point about Tyler's household situation is worthy of mention. With the help of an unmarried sister, her daughters, and hired girls, Tyler embarked on an ambitious plan for producing homespun cloth. Initially learning to spin flax and cotton after her father's business failure forced the family to relocate from Boston to Framingham, Tyler put these skills to use in her own household to produce thread that she then gave to a neighbor to weave. This initial reliance on female networks of neighbors and relatives to produce cloth might also have made it possible for information about child care to pass among several generations of women, a circumstance elided in Tyler's textual representation of her maternal authority as based mainly on her own experience. In the decades before she wrote *The Maternal Physician*, however, Tyler decided it was more cost-effective to purchase a loom and use the labor within her household, supplemented by additional hired help. She used this home-produced cloth for the household as well as for the children's everyday clothing. As Laurel Ulrich notes, Tyler and others were intensifying a colonial production system in the early decades of the nineteenth century. Although the manufacture of cloth reflected colonial ideals of linen-centered cleanliness, it overlapped with a different regime for health and cleanliness, outlined in *The Maternal Physician*, which called for the daily bathing of children.[17]

Tyler's family history, educational background, and marriage to a prominent writer and judge help us to identify the politics of her authorship and her medical advice. She took the idea of republican motherhood to its literal and logical conclusion—the production and nurturing of potential citizens' bodies. In so doing she provided an American translation for the British imperial concern with improving maternal care for infants, what one scholar has described as "colonizing the breast." Tyler's identification with the fate of the new nation appeared as early as the dedication page of *The Maternal Physician*, where she explicitly connected her mother's relationship to her helpless child (Tyler) with women's duties to nurture the new nation.[18]

Tyler claimed to have been motivated to write her guide for reasons similar to the imperial concerns that motivated English physicians Hugh Smith and Michael Underwood: when reading some "old" newspapers, she was struck by the fact that so many children died before reaching the age of two. She laid the blame for this mortality squarely at the door of ill-informed mothers and nurses, who, she believed, mismanaged children to death, either directly, with bad care, or indirectly, by too quickly consigning them to the care of physicians. This was an interesting twist on the receipt book's goals of independence from doctors who were as likely to kill as to cure. In contrast, when she considered the blooming health of her own growing brood of children (who, her reasoning implied, had not been mismanaged) she could only look heavenward and give thanks. Her record as a mother became part of her apologia for putting *The Maternal Physician* before the public: "Eight lovely and beloved children, who have all (except the three youngest) passed through the usual epidemics of our country, and now enjoy an unusual proportion of health and strength, are the best apologies I can offer for thus presuming to give my advice unasked, and perhaps undesired, to my fair countrywomen" (17). With her fellow American matrons as her intended audience, Tyler proposed nothing less than to "take the babe from birth, and attend it through every stage until it is two years old; after which period children . . . will increase in health and strength without any attention except the ordinary care conducive to cleanliness and exercise, two points never to be dispensed with through life" (18).

Tyler organized her book around the assumption "Every Mother her Child's best Physician" (5), a clever play on the stock claim of several popular medical manuals, "Every man his own physician" or "Every man his own best physician." In the early pages of her book, she modestly distanced herself from her own advice, noting that her frequent summons of a doctor for the minor illnesses of her first child had provoked him to advise her, "[Y]ou may yourself be your child's best physician . . . if you only will attend to a few general directions" (6). In other words, a doctor legitimated the claim "Every mother her child's best physician." Dutifully following this advice, she told her readers, she met with much success. As she explained to her readers how gratitude to this physician had impelled her to write, she transformed her intellectual debt to him into something all her own: "[W]hy then may I not show my gratitude by presenting to the matrons of my country the fruits of *my* experience, in the pleasing hope that I may be instrumental in directing them in the all-important and delightful task of nursing those sweet pledges of connubial love" (6, emphasis mine).

Whatever the actual origin of the idea that a mother might be the best physician to her child, Tyler clearly saw her book as both the product of her own ex-

perience and a tool to assist other American matrons to become more effective
caretakers of their children. Again, she was formulaically apologetic: "The motive, I trust will ensure my pardon for any traits of egotism, which must un-
avoidably appear while recommending a mode of treatment founded chiefly
upon my own experience, but which, nevertheless, the better to enforce, I intend to enrich with casual extracts from the most improved medical authorities" (18). "Improved" medical authorities (which became "approved" medical
authorities in the second edition) thus enforced but did not ground her system.
Tyler carefully maintained her intellectual independence from the other toilers
in the advice book genre, all of whom happened to be male.

This combination of seeming deference to medical authority and confident
self-assertion—a variant of Paschall's independent judgment—was repeated
throughout Tyler's volume. Although she acknowledged that she had learned
"many useful hints" from the writings of "the most able" gentlemen physicians,
she insisted that in most ordinary circumstances a mother knew her child's
body—and cared about it—better than anyone: "Who but a mother can possibly feel interest enough in a helpless new born babe to pay it that unwearied,
uninterrupted attention necessary to detect in season any latent symptoms of
disease lurking in its tender frame, and which, if neglected, or injudiciously
treated at first, might in a few hours baffle the physician's skill, and consign it
to the grave" (7).

Although Tyler's claims might seem to have sprung from a belief in women's
innate aptitude for taking care of children, in fact, they were contingent upon
the mother's ability to "observe all the minutiae of her child's state of health"—
for example, body temperature, appetite, coloring, as well as the early signs of
illness—while nursing (9). In this, she called upon the growing body of advice
literature, most of it British in origin (Smith and Underwood), urging women
to nurse their own infants for reasons of health rather than putting them out
to wet nurses. If women were to cast off this "sweetest privilege of nature,"
Tyler averred, her system would "fall to the ground" (9). Seeing breast-feeding
as an opportunity for mothers to gather data on healthy children, however, represented a significant departure from male advice book authors, who described
nursing mothers passively offering a breast to infants and noting details about
the child's health only when it was sick.

In emphasizing the mother's relationship with her child and in eschewing
customary methods of childrearing, Tyler reflected the influence of Enlightenment theories of child development. Advocating greater physical freedom for

the developing infant so that it could follow the path dictated by nature rather than by wrong-headed custom, doctors such as William Cadogan, Bernhard Christoph Faust, A. F. M. Willich, Buchan, and Smith advised dispensing with confining clothes, swaddling bands, and stays. In keeping with this back-to-nature, rational approach to childrearing and highlighting her own modernity, Tyler similarly rejected the traditional explanations and confining devices resorted to by "great grandmammas" (136). For example, she classified left-handedness as a naturally occurring defect rather than as the fault of the mother (128–29). She also disputed that early walking indicated superior health, urging parents not to use artificial means to force children to walk before they were ready. Creeping freely on a carpet, a child would develop a stronger and more elegant form than those either confined to walking stools or forced to walk too soon (132–36).[19]

Echoing the advice of other Enlightenment doctors, Tyler's system centered on a regime of cleanliness, which included daily cold baths for newborns, daily changes of clothes and bed linens, and fresh air. Tyler advocated applying water from a cold basin with a warm hand rather than immersing the child, after which the "nurse" should wipe it "perfectly dry" with a "warm soft cloth" (22). "This washing I would have repeated every morning," she instructed. The point of this daily bath was to keep the baby's skin healthy and prevent "excoriation," or sores and chafing. Anticipating her readers' fears that cold bathing was unnecessarily cruel or unnatural for infants, she replied, "Now I would ask, which is the most cruel or unnatural; to lave its little limbs with the pure element designed by a beneficient Creator for our purification, and consequent health, and beauty," or to treat the skin "already perhaps in many places excoriated" with ardent spirits that "must occasion intolerable smarting and pain" (23). Tyler urged her readers to experiment with both methods to see which one made the child cry less (23). In suggesting that mothers experiment, she parted company with Underwood, who condemned the regime of cold water immersion that so discomfitted tender-hearted mothers. Hers was a practical assessment in which infant suffering might be rationally calculated and weighed against the benefits of the treatment.[20]

In addition to the daily bath, Tyler insisted on frequent changes of clothing. Like the medical authors she had read, Tyler emphasized clean linens because of fears that soiled linens would allow the "little frame, already loaded with disease, to imbibe again the bad humours kind nature is struggling to expel from the innumerable pores of the skin" (180). She was unique, however, in observing that changing clothes allowed a dutiful mother additional opportunities to gather data on her child's body.

Throughout the daily bathing and changing of young children, Tyler emphasized the tactile role of the mother or nurse. Women removed children's clothing, exposing their bodies to view. They used their hands to apply water all over the child's body, including the genitals, and then dried the entire body with a towel. Although nothing in Tyler's recommendations suggested that such practices could be sensual, her directives fostered physical intimacy and tactile bonds between mother and child, perhaps laying the groundwork for the bourgeois supervision of children's bodies that Foucault claimed gave rise to modern sexuality.[21]

Along with several adaptations of Faust's *Catechism of Health*, Tyler's *Maternal Physician* was among the first American medical advice books to urge readers to adopt daily baths for their children as the best protection for their health. Although friction remained an important part of the cleanliness regimes these books advocated, water had become increasingly important, as a means of both stimulating the circulation and cleansing the skin. The two books also stressed the tactile, sensual presence of the mother or nurse at bath time: both books recommended touching the baby's wet skin with the hand. In Tyler's book, bathing and frequent changing of clothing also provided the mother with additional information about her child's body while it was *in health*, an important departure from the receipt book emphasis on gathering data about sick bodies as well as previous guides. This knowledge, combined with that gathered during breast-feeding, was the foundation of the American matron's medical authority and the strongest support for her claim to be her child's best physician.[22]

What about Tyler's specific textual strategies? Tyler sprinkled poetry liberally throughout her book. The poetry of Luigi Tansillo's "The Nurse," translated by William Roscoe, suffused the entire work with a sentimental appeal to women to breast-feed their own children:

> Not half a mother she, whose pride denies
> The streaming beverage to her infant's cries,
> Admits another in her rights to share,
> Or trusts its nurture to a stranger's care.   (10)

Buttressed by Tansillo, Tyler offered a highly romantic view of the joys and inherent beauty of breast-feeding. "Clasp your fair nurslings to your breasts of snow / And give the sweet salubrious streams to flow," she quoted (46). "The starting beverage meets the thirsty lip. / Tis joy to yield it, and 'tis joy to sip" (30). In addition to using sentiment to advocate militantly for breast-feeding, Tyler's selections from Tansillo reassured readers that this duty of motherhood would not compromise their gentility, symbolized by the whiteness of the snowy breast.[23]

Her quotes from another source, John Armstrong's poem *Art of Preserving Health*, appear to have had three purposes: to offer coded revelations about her own life situation, to hold the interest of her female readers with flowery verse, and to present technical information that might otherwise be deemed dry. She quoted Armstrong on the inviting rural wilds of nature immediately before citing her own record as a mother—eight children born, eight still living—thus echoing the wisdom of other medical guides that the rigors of rural life offered the healthiest environment for children. But it also quietly revealed her own situation in rural Vermont, hinting that she saw this as part of her good fortune as a mother. Armstrong's verses also neatly explained in three lines how the effusions of the skin could cause illness, a topic that had inspired eighteenth-century English doctor George Cheyne to write enthusiastically for several dozen pages:

> The grand discharge, th'effusion of the skin
> Slowly impaired, the languid maladies
> creep on, and through the sinking functions steal.[24]

As we have seen, Tyler couched her advice in terms of her own experience as the mother of eight living children. She thus had the benefit of appealing to the American reader's interest in an American mother's experience—an important attraction in a new nation that saw the unparalleled virtue of its mothers as necessary to sustain its experiment with republican forms of government and already believed in the cultural distinctiveness it was working so hard to establish. But she also called upon prominent medical authors for corroboration, which allowed her to embrace the popular health manual's tactic of recommending advice on regimen that readers were predisposed to accept as common sense.

Tyler did not simply provide random extracts from medical writers, however. She selected only certain writers—Armstrong, Buchan, Smith, and Underwood—and she evaluated their advice against her own experience. Many other notable authors—Tissot, for instance—went uncited. More than simply seeking reinforcement and corroboration, however, Tyler vetted the competing advice of some of the most well-known medical advice book authors of the day and tried to craft a consensus within her text about a mother's best course of action. The best example of this is when she compared advice about giving a child solid food before weaning. After tabulating the opinions (Smith and George Wallis were for it, Buchan advised waiting until the infant was four months old, Underwood and Thompson thought it mattered little), Tyler sided with the latter two authorities based on her experience with her own children, who followed different sequences but were all weaned without incident (142). A second purpose was to act as a conduit of knowledge to women less learned than herself: "I will now resume my quotations," she declared after an apology for rambling, "well pleased to present to those mothers who cannot conveniently obtain the work [of Michael Underwood] itself, the fruits of so much real humanity, learning and experience" (53).

Tyler was not always such an uncritical advocate of academic medicine. She viewed the distinction between medical theory and folk belief with skepticism, as in her lengthy defense of the claim that breast milk will dry up if it is thrown away rather than fed to an infant. The fact of unused milk causing a breast to dry up could not be accounted for by "any known principles," she admitted, but she insisted, "[F]acts, however, are not made by theory, but theory created by facts." If a natural phenomenon like the drying up of breast milk "is noticed by the vulgar" but unexplainable to the learned, she contended, the learned deride those who believe it as weak and credulous. Once the fact is confirmed and the learned devote some effort to explaining it, "they publish their reveries; and this is called theory." Those who fail to admit the fact and the theory are then "ranked among the ignorant vulgar" (13). Having dismissed the significance of this divide, Tyler felt free to side with the ignorant vulgar. But, as her book's conclusion reveals, she also sought the approval of regular physicians, whom she invited to read her book and offer corrections, reassuring them that any errors were the result of ignorance and inexperience, not vanity or empiricism (283).

Tyler did not hesitate to modify or correct the advice offered by even favorite medical authorities. Of Buchan's suggestion that a singed rag or raisin be applied to the umbilical cord stump, followed by an application of alum water or sugar of lead, Tyler noted coolly, "I have always been in the habit of roasting the raisin, and grating upon it a little nutmeg; and I think the dressing ought to be renewed every day, and the part anointed with sweet oil in preference to the waters recommended above" (42). She urged her readers to consult Underwood for more particulars but then acknowledged that some readers, who "may not have it either in their power or inclination to consult so large a work," would appreciate her inclusion of extracts (43). Again, after noting Underwood's recommendation of cabbage leaves to deal with excess navel discharge, Tyler added that Turner's cerate had worked best for her (44). She also quibbled with Underwood's explanation that infant flatulence was due to taking inappropriate food, noting that children fed only breast milk also could be afflicted. She urged women to watch their diets while breast-feeding, noting that her own children had had trouble with gas when she had eaten apples (67).

In parting company with the popular advice literature, much of which was English, Tyler occasionally positioned herself as a cultural translator who could help the American reader find native medicines and remedies equivalent to those recommended in European texts. In one such instance, she informed her readers: "[O]ur country boasts, among a vast variety of vegetable productions, a plant generally known by the name of dragon root or wild turnip, that is an excellent remedy in this complaint, and for the canker in all its forms. It is used in the same manner Dr. Underwood prescribes Borax" (59). Tyler admitted limits to her ability to explain why these substances worked similarly: "I am not scientific enough to describe technically the virtues of this root; but I know that when fresh it resembles a common English turnip in its appearance, [and] is of a smart pungent taste" (59). Neatly illustrating the importance of efficacy for her regime, she then offered a description of the root's performance as an emetic. In similar fashion, her list of herbs and medicinals drew mainly on plants native to New England. She warmly recommended blood root to the gentlemen of the faculty as an overlooked but effective herbal remedy (259).

Tyler's militant advocacy of breast-feeding also occasionally led her to take issue with more lenient male authors. Thus, for example, she roundly criticized Hugh Smith for his suggestion that women could regulate the feedings of infants to minimize the fatigue that comes with nursing a child. Although Tyler conceded that a feeding schedule might not harm a child, she found Smith's uncritical acceptance of the idea of female fragility annoying: "I have often been vexed with physicians who, while they exhort us to follow nature, from a misplaced indulgence to the prevailing fastidiousness of the age, adopt the absurd notion that a mother cannot endure the fatigue of suckling her own child" (35). Tyler revealed her own view of female nature as basically hardy but swayed by fashion to neglect maternal duty: "Such unhappy mothers are to be pitied, but I greatly fear the far greater number who neglect this sweet endearing office are more fit objects of censure than of pity" (35). Educated women whose self-indulgence weakened their own health, she found the most guilty of neglecting this maternal duty.

Tyler enjoined "every intelligent mother to improve her judgment by consulting the most approved medical books on the treatment of children, but more especially by observation and attention to their constitutions and complaints" (73). This combination of book learning and observation would enable them "to judge when nature really requires assistance, and how to administer it with propriety, in all *common complaints*" (73). Tyler did recognize limits to maternal medical authority, however, and cautioned her readers not to make the mistake of assuming they could cure all their children's ills. She thus participated in delineating the boundaries between domestic healing and professional medicine: "If your infants are really ill, so that their complaints will not readily yield to the common palliatives, or the often greater efficacy of good nursing, let me entreat you not to delay calling in professional aid until your children are too far gone to admit relief. Thousands of helpless little innocents suffer greatly from the too prevalent opinion (especially among the lower classes of society) that a physician cannot tell what to do for such young children" (73).

As is probably evident, Tyler's advice, like that of Tissot and Buchan before her, contained a vision of class differences, mediated by charitably minded, educated women. In addition to calling attention to the particular sufferings of the children of the poor, due to false assumptions about the inappropriateness of doctors to tend to infants or out of a mistaken belief in a harmful folk remedy, Tyler also envisioned her female readers being able to minister to the needs of poor neighbors: "Every lady who has a family, or who wishes to impart the blessings of her medical knowledge to her poor neighbours, should furnish herself with a medicine chest, containing every drug of known and established efficacy, and a set of scales and weights for the purpose; for no one but a regularly bred physician should ever venture to give any potent drug, especially opium, calomel, or emetic tartar, unless weighed with scrupulous exactness, according to the above-mentioned rules" (175).

But Tyler's vision of class relations was not confined to educated women acting as physicians to the poor. In addition to critiquing educated, fashionable women for failing to breast-feed, she also condemned them for leaving their children in the hands of hired women who did not know or care enough to provide children with the care they needed. Tyler's recommendations for nursing and cleanliness urged mothers to take active roles in collecting data about their children's bodies and making informed judgments about their health based on hands-on experience. Her regime admitted only the strictly supervised nurse and defined the mother's role as one of continuous physical and intellectual connection to her child's body, both in sickness and in health.

There are a number of ways we might assess the significance of Mary Tyler and her singular foray into print culture. Using the broadest brush (and a Foucauldian one at that), we might see her as an agent in the creation of the bourgeois body, whose regimes of discipline and cleanliness testified to and made possible broader deployments of power and knowledge in a new regime of sexuality and self-care. That she was a woman writing makes little difference in this analysis, although her claims as a mother might encourage us to revise Foucault to take some of the focus off the family writ large and put it onto the evolving relationship between mother and child. In particular, we might reconsider the centrality of the bourgeois mother's powerful sensual connection to and extraction of information from her child's body.

If we treat *The Maternal Physician* as a window onto women's changing relationship to medical authority—including their changing embodiment of that authority—different analytic possibilities come into view. *The Maternal Physician* gives a different impression of the early nineteenth century than that gained from Laurel Ulrich's *A Midwife's Tale*. In the world of Ulrich's Martha Ballard, the day of the midwife-healer is passing and the popularity of the male physician is rising, even in the birthing chambers of rural Maine. Wedded to the traditions of midwifery and herbal healing, midwives like Ballard could not embody the promises of modernity to their female clients the way an academically trained man with forceps could. Like making homespun cloth and hiring out young adults to neighbors, traditional midwifery, as practiced by Ballard, gave way to newer methods and fashions.

Tyler's book offers a more optimistic view of the opportunities republican motherhood opened up for women. These included limited medical authority over their children and new responsibilities for producing healthy bodies as well as educating the minds of male citizens. Tyler takes this responsibility seriously; she candidly explains the stakes to her readers: "As for your sons, let me entreat you to reflect upon what manner of men you will wish to see them in after life . . . Do you wish to see them effeminate and pusillanimous, then be it your care to guard their complexions, to instil in their tender minds the love of dress and show, to lead their attention to the best drest guest, and most splendid equipage . . . But if, on the other hand, you wish to rear the hero and the sage, teach them betimes to set no more than their just value on the trappings of fashion" (280–81). She is asking, in essence, whether they wish to be the mothers of Colonel Manly or of Dimple, a question loaded with political baggage. In the answer, an affirmation of a uniquely American male character and a rejection of European ways, Tyler carves out her niche in the genre of medical advice literature, previously dominated by European male doctors. She is bound to treat the health needs of American children as distinctive because such a national claim justifies her own literary effort. The child she is preparing for the duties of male citizenship, moreover, has been given daily baths and clean clothes, and has had his body closely scrutinized, even touched, by his mother or a hired nurse. Yet he has not become soft or effeminate as a consequence. A mother's intimate physical care for a son's body, according to Tyler, when combined with the appropriate message about the follies of fashion, simultaneously nurtures his manhood and expands her domestic authority at a time when medical men were continually making inroads into female healing traditions.

The maternal authority Tyler constructs for her readers is circumscribed by the superior learning of trained physicians but reigns supreme in the ordinary course of childhood health and illness. It rests, moreover, on the mother's intimate supervision of healthy young bodies and on the data collected from regular breast-feeding and bathing. It would be years before any other American woman dared to articulate a claim to medical authority based on her own experience. One of those who eventually did, Catharine Beecher, wrote numerous treatises on health and household management, directed at married female readers. Unlike Tyler, however, Beecher could not rely so heavily on her own experience, at least not for her advice about health. Having never been a mother, she resorted frequently to one of Tyler's textual strategies—that of being a translator of foreign and learned advice for a more homespun American readership. Ironically, when we think of an exemplar of the saying "Mother knows best," a close paraphrase of Tyler's boldest claim, we think of Beecher as the high priestess of American domesticity and as the female reformer who did the most to build women's authority within the home. Perhaps a more accurate embodiment of the phrase is Mary Tyler, who strove to link the claims and content of republican motherhood with a scientific effort to preserve the lives of future citizens and to negotiate an expanded maternal authority.

NOTES

This essay benefited immensely from the helpful suggestions of Charles Rosenberg and from the insights of the following audiences: the University of Pennsylvania History and Sociology of Science Seminar, the Lehigh University History of Technology Seminar, the Penn Women's Studies Feminist Theory Seminar, the McNeil Center for Early American Studies Brown-Bag Series, the University of Michigan Women's Studies Lecture Series, and the University of Florida Department of History Lecture Series.

1. *The Maternal Physician: A Treatise on the Nurture and Management of Infants, from the Birth until Two Years Old. Being the Result of Sixteen Years' Experience in the Nursery. Illustrated by Extracts from the Most Approved Medical Authors* (New York: Isaac Riley, 1811), 174–75; *The Maternal Physician* (Philadelphia: 1818). Subsequent references to *The Maternal Physician* are from the 1811 edition and appear parenthetically in the text (they are the only references treated this way in the chapter). Child noted: "The author of the Maternal Physician states a curious fact. He says when the process of nursing is very painful to the mother, the milk is sometimes drawn out with sucking-glasses; if the child is fed with it, a supply will remain in the breast some time; whereas, if it is thrown away, it will gradually diminish till it ceases." Child, *The Family Nurse* (Boston: Charles J. Hendee, 1837), 36.

2. By the early nineteenth century, discussions of pregnancy and childbirth—midwives' terrain in the seventeenth century—were dominated by scientific treatises on female anatomy. For an example of seventeenth-century midwives' authority in the birthing chamber, see Jane Sharp, *The Midwives Book Or the whole Art of Midwifry Discovered* (London: 1671). See also Adrian Wilson, *The Making of Man-Midwifery: Childbirth in England, 1660–1770* (Cambridge: Harvard University Press, 1995), 1–2, 6, for the growing number of male authors of midwifery books in the eighteenth century, and Judith W. Leavitt, *Brought to Bed: Childbearing in America, 1750–1950* (New York: Oxford University Press, 1986), for the changing situation in the United States.

3. Hugh Smith, *Letters to Married Women on Nursing and the Management of Children* (Philadelphia: 1792), vi; Michael Underwood, *A Treatise on the Diseases of Children, and Management of Infants From the Birth* (Boston: 1806), 430.

4. See, for example, George Cheyne's claim that he published at the request of a friend for whom he had originally composed his rules of health: Cheyne, *An Essay of Health and Long Life* (London: George Strahan, 1724), xi–xii. In the preface to *The Catechism of Health: Containing Simple and Easy Rules and Directions for the management of Children, and observations on the conduct of health in general* (New York: 1819), William Mavor claimed that his shorter version of Bernhard Christophe Faust's *Catechism of Health* (New York: 1798) had been adapted to the situation in the United States. Faust's manual was first published in German in 1792 and in English in 1794. The first American edition appeared as *The Catechism of Health: Selected from the German of Dr. Faust; and considerably improved by Dr. Gregory, of Edinburgh* (New York: 1798).

5. See for example, Rosemary Zagarri, *A Woman's Dilemma* (Wheeling, Ill.: Harlan Davidson, 1995), 132–49, for her discussion of Mercy Otis Warren's strategies for legitimating her female perspective in her history of the American Revolution. See also Nina Baym, "Mercy Otis Warren's Gendered Melodrama of Revolution," *South Atlantic Quarterly* 90 (summer 1991): 531–54.

6. Rousseau, *Emile or On Education*, ed. Allan Bloom (New York: HarperCollins, 1979), 44–49, quotation on 46; Tissot, *Advice to the People in General, with Regard to their Health; But particularly calculated for those, who are the most unlikely to be provided in Time with the best Assistance, in acute Diseases, or upon any inward or outward Accident* (London: 1767), 15; William Buchan, *Domestic Medicine: Or a Treatise on the Prevention and Cure of Diseases by Regimen and Simple Medicines* (Philadelphia: 1784). For the popularity of Buchan's work, see Kevin J. Hayes, *A Colonial Woman's Bookshelf* (Knoxville: University of Tennessee Press, 1996), 83–100, quotation on 95; Charles Rosenberg, "Medical Text and Social Context: Explaining William Buchan's *Domestic Medicine*," in *Explaining Epidemics and Other Studies in the History of Medicine* (Cambridge: Cambridge University Press, 1992), 32–56.

7. Many of the published cookbooks literate colonial women were likely to have on their bookshelves borrowed heavily from the receipt book genre. See Hayes, *Colonial Woman's Bookshelf*.

8. Elizabeth Coates Paschall, receipt book, 10, Library of the College of Physicians of Philadelphia; quoted from with permission of the Library. See Ellen Gartrell, "Women Healers and Domestic Remedies in 18th Century America: The Recipe Book of Elizabeth Coates Paschall," *New York State Journal of Medicine* 87 (1987): 23–39.

9. Paschall, receipt book, 3, 7.

10. Robert Boyle, *The Philosophical Works of the Honourable Robert Boyle, abridged, methodized, and disposed under the general heads of physics, statics, pneumatics, natural history, chymistry, and medicine. The whole illustrated with notes, containing the improvements made in the second parts of natural and experimental knowledge since his time*, 2d ed., corrected, 3 vols. (London: W. Innys and R. Manby, 1738), 3:661.

11. Harold Donaldson Eberlein, "When Society First Took a Bath," in *Sickness and Health in America: Readings in the History of Medicine and Public Health*, ed. Judith Walzer Leavitt and Ronald L. Numbers (Madison: University of Wisconsin Press, 1978), 331–41; Richard L. Bushman and Claudia L. Bushman, "The Early History of Cleanliness in America," *Journal of American History* 74 (March 1988): 1213–38.

12. Apparently, many ordinary men believed it possible for a man to be too well coifed, too well dressed, and, indeed, too clean, even as others eagerly took advantage of their freedom to dress like men of leisure and privilege. From as early as the 1770s, cartoons, rhymes, and satiric plays made fun of men who appeared to be overly focused on their own persons. Lampooning macaronis, fops, dandies, mollies, and men of fashion— known in France as les petits maîtres—broadside and pamphlet authors and engravers questioned the manhood of men devoted to grooming, fine clothing, and appearance. Cartoon dandies became feminized, exhibiting swollen hips and breasts and tiny waists. One example contemporary with Tyler's second edition of *The Maternal Physician* was Buzz Bumblery, *Ephemera, or the History of Cockney Dandies* (Philadelphia: 1819), a satirical poem with illustrations of Philadelphia's dandies, whom the author deemed to be pale imitations of the London original.

13. This reaction against men of fashion lingered well beyond the early republic, suggesting the persistence of male interest in dressing "out of season" in both a class and gender sense, and offering hints of a possible generational divide. Novels and children's stories drew pointed morals about young men who ruined themselves with lavish expenditures on fancy clothes. Such literary lessons even make rare appearances in female diarists' accounts of male dress.

14. In one example from a Boston edition of an English work on adultery and divorce, *Cuckold's Chronicle: being select trials for adultry, incest, imbecility, ravishment, etc.*, vol. 1 (Boston: 1798), the engraving "Criminal Conversation" showed a man (presumably the aristocratic woman's adulterous lover) peeping at her through a window in the bath house as she is about to take her bath.

15. In the short term, Rush was unsuccessful, although he persisted in his quest through at least 1810, when Abigail Adams wrote to him approvingly about jettisoning Latin and Greek as a means of creating an even educational playing field for men and women. See Linda Kerber, *Women of the Republic: Intellect and Ideology in Revolutionary America* (Chapel Hill: University of North Carolina Press, 1980), 218–21.

16. Frederick Tupper and Helen Tyler Brown, eds., *Grandmother Tyler's Book: The Recollections of Mary Palmer Tyler (Mrs. Royall Tyler), 1775–1866* (New York: G. P. Putnam's Sons, 1925), 130, 135.

17. Ulrich, "Wheels, Looms, and the Gender Division of Labor in Eighteenth Century New England," *William and Mary Quarterly*, 3d ser., 55 (1998): 3–38.

18. Ruth Perry, "Colonizing the Breast: Sexuality and Maternity in Eighteenth-Century England," *Journal of the History of Sexuality* 2 (March 1991): 204–34.

19. Faust, *Catechism of Health*; A.F.M. Willich, *Lectures on Diet and Regimen: being a systematic inquiry into the most rational means of preserving health and prolonging life: together with physiological and chemical explanations, calculated chiefly for the use of families, in order to banish the prevailing abuses and prejudices in medicine*, 2d ed. (London: 1799).

20. Underwood, *Treatise on the Diseases of Children, and Management of Infants From the Birth* (Boston: 1806).

21. Michel Foucault, *The History of Sexuality I, an Introduction*, trans. Robert Hurley (New York: 1980).

22. *A New Guide to Health compiled from the Catechism of Dr. Faust; with additions and improvements, selected from the writings of medical men of eminence* (Newburyport, Mass.: W. and J. Gilman, 1810). Unlike the first American edition of Faust cited above, this adaptation eschewed the religious and biblical justifications at the beginning to focus on the state's need for a robust citizenry. See also Charles Rosenberg, "Catechisms of Health: The Body in the Prebellum Classroom," *Bulletin of the History of Medicine* 69 (1995): 175–97.

23. Kim F. Hall, *Things of Darkness: Economies of Race and Gender in Early Modern England* (Ithaca, N.Y.: Cornell University Press, 1995).

24. John Armstrong, *The Art of Preserving Health: A Poem* (Boston: Green and Russell, 1757), 28.

5

THOMAS A. HORROCKS

Rules, Remedies, and Regimens

Health Advice in Early American Almanacs

Marion Barber Stowell, a historian of early American almanacs, wrote in 1983 that modern versions of the genre were the "degenerate offspring of respected ancestors whose contents were not primarily advertisements for hair-growing and itch-relieving potions." Almost a century earlier, another historian of almanacs, Samuel Briggs, expressed a similar opinion. Briggs lamented that the almanac had degenerated into a publication whose columns teemed "with the virtues of pills, potions and plasters, interspersed with views of our internal economy calculated to make the well man ill, and the invalid to relax his grasp on the thread of life."[1] Stowell's assessment of almanacs in the 1980s is exaggeration, but Briggs's assessment of the genre in the late nineteenth century is less so. By the time Briggs was issuing his indictment, the almanac trade had fallen under the domination of patent medicine and pharmaceutical companies.

Briggs and Stowell wrote wistfully of a world that had been lost—a pre–Civil War world in which almanacs offered their readers not various brands of hair-growing and itch-relieving nostrums, but advice, enlightenment, and entertainment. It was a world in which Poor Richard, Poor Robin, Isaac Bickerstaff, Abraham Weatherwise, and the Old Farmer passed on farming tips, anecdotes, astrological information, humorous stories, maxims, weather predictions, and health advice. Because the almanac of this world offered something to almost everyone and almost everyone could afford to purchase one, it was, in the words of Briggs, an "honored guest at every fireside."[2]

Briggs and Stowell—as well as others—have stressed the almanac's importance as a valuable resource for the study of popular culture in America from the colonial period to the Civil War.[3] While I disagree with Briggs's and Stowell's apparent dismissals of later almanacs—specifically patent medicine almanacs—as unworthy of scholarly attention, I concur with their shared belief in the historical value of those published before 1860. The almanac, more than any other genre of print (with the possible exception of the newspaper), serves as a lens through which twenty-first-century scholars can examine American popular culture of the seventeenth and eighteenth centuries and the first half of the nineteenth.[4] Almanacs are indispensable resources for the study of popular medicine in early America. This chapter is part of a larger study of health advice in pre-1860 American almanacs and what that advice tells us about popular views of the body, health, and disease in early America. It examines the three main forms of health advice presented in these publications: astrological rules, remedies for various ailments, and regimens for preserving health and long life.[5]

The genre already had a long history by 1639, the year this country's first almanac was printed in Cambridge, Massachusetts. The almanac in its seventeenth-century guise was a composite of three chronological devices that dated back to antiquity. British and European almanacs included a *calendar,* or a listing of days of the week and months that noted church festivals and saint's days; an *almanac,* which consisted of astronomical and astrological compilations of the passage of time; and a *prognostication,* which consisted of astrological predictions of political and social events. America's first almanacs, sometimes referred to as "Cambridge" or "Harvard" almanacs because they were compiled by Harvard graduates or graduate students and printed in Cambridge by a press that was controlled by the college, resembled English almanacs in certain ways. Like their British counterparts, most American almanacs issued during the seventeenth century included a preface, a calendar, information on eclipses, and a listing of local court and fair dates.[6]

There were significant differences, however, between Harvard and English almanacs. The former, compiled by staunch supporters of Puritan orthodoxy, replaced saint's and feast days with important historical dates and identified months with numbers rather than with traditional names, which were considered pagan in origin. Most important, English almanacs devoted much attention to astrology while Harvard almanacs emphasized astronomy, learning, and Puritan doctrine over what their compilers considered heathenish prognostications of health, the weather, or both. And while their English counterparts sought to entertain a general audience of readers for profit, Harvard almanacs sought to edify and instruct an audience of readers as well educated as their compilers.[7]

By the last decades of the seventeenth century, American almanacs were beginning to offer their readers more variety, and Harvard almanac makers were losing their hold on the market to competitors who correctly sensed that almanac readers wanted more predictions, less piety, and some levity along with their learning. And readers wanted astrology. By the end of the century the almanac had assumed the basic form with which Americans of the eighteenth century were familiar: a publication consisting of a calendar, astronomical and astrological compilations of time, and weather predictions.[8] During the eighteenth century other elements were added to the almanac, such as extracts from literature, poetry, lists of roads, local and federal court dates, postage and currency information, farming advice, humor, and health information. The almanac had become the most popular secular publication in America by this century's end. For many Americans it was a fount of information and entertainment; a repository of vital facts for the farmer, merchant, and seaman; and a place where one could seek respite from the unchanging routines of everyday life.[9]

The format that almanacs assumed by the beginning of the eighteenth century remained consistent throughout the century and well into the nineteenth. Ranging in length from twenty-four to thirty-six pages (several nineteenth-century almanacs ran to forty-eight pages, and almanacs that also served as city or state directories contained more than one hundred pages), almanacs generally included three main sections. The first section, or front matter, comprised a preface or introductory statement from the almanac maker; eclipses for the coming year; names and symbols of the planets; the five aspects;[10] the twelve signs of the zodiac, the "Anatomical Man," or both;[11] and other astronomical and astrological information.[12] Occasionally, this first section would contain an explanation of the astronomical and astrological symbols and terms that appeared in the almanac. The second section contained the calendar, which was divided into seven or eight columns for each month. The columns listed the days of the month; times of the rising and setting of the sun and moon; times of high tide; the moon's place in the zodiac; feast days and other historical dates; and weather predictions. The calendar pages also included the phases of the moon for each month, poetry, and sometimes maxims or proverbs, as well as short essays on farming, health, and the virtues of honesty, morality, and thrift. The third section comprised interest or currency tables; local and federal court schedules; a listing of local roads, which included the distance between various towns; essays on various subjects; humor; anecdotes; and, beginning around the 1830s, patent medicine advertisements. Generally, the reader would find farming tips and health advice in this section.

Low production costs and a ready market made almanac production a profitable venture for many colonial printers.[13] Affordable prices (ranging from six to twelve cents per issue during the first half of the nineteenth century) made almanacs accessible to Americans at the lower end of the economic scale, although it must be stressed that almanacs circulated widely in—and were popular with—all economic classes.[14] In his *Autobiography*, Benjamin Franklin referred to his *Poor Richard* almanac "as a proper Vehicle for conveying Instruction among the common people." Nathaniel Low claimed that he did not design his publication "to inform the learned" but rather for the "few Poor and Illiterate" who were "not so biased against the arts as the multitude are." Nathaniel Ames Sr. was even more blunt when writing about his almanac's intended audience: "[T]his Sheet enters the solitary Dwellings of the Poor & Illiterate, where the studied Ingenuity of the Learned Writer never comes."[15] Passages such as these, together with the almanacs' affordability and small size (most could fit into a person's pocket), should not mislead us into thinking that these publications were issued solely for the poor and semiliterate. Many almanacs contained philosophical essays, extracts from medical and scientific treatises, references to classical literature, poetry (including the works of Dryden and Pope), and selected passages from contemporary works of fiction, which suggests that early American almanac makers were targeting the widest possible audience.[16]

By attracting a readership comprising men and women, the affluent and the poor, the learned and the semiliterate, the artisan and the farmer, and the urban as well as the country resident, almanac makers, like American printers in general, were in a way acting as cultural mediators. Historians have shown that, generally, early American almanac makers and printers came from the artisan class—the same strata as surveyors and ship captains. Literate but not highly educated, almanac makers, to be successful, had to know the literary tastes of readers from different social and economic groups. In short, almanac makers had to select material for their publications that would meet the needs of a diverse readership, ranging from the affluent, learned reader to the poor, semiliterate one. At the same time, they were introducing readers from different social and economic spheres to material that they might not usually have encountered. That many almanac makers were successful in attracting such a broad readership is a prime reason for the genre's popularity in the second half of the eighteenth century and the first half of the nineteenth.[17]

The public's growing demand for almanacs and printers' desire to make profitable products led to a proliferation of almanacs during the end of the eighteenth century and early decades of the nineteenth.[18] But as successful as the almanac trade was, it was not immune to pressures from within the publishing community in particular or American society in general. While almanac makers competed with each other, the almanac itself faced increasing competition in the nineteenth century from other forms of popular print and the growing numbers of newspapers and magazines.[19] Moreover, the almanac's traditional roles as calendar and keeper of time were increasingly challenged in the middle decades of the nineteenth century by the emergence of advertising calendars, clocks, and inexpensive watches. In 1786, almanac maker Nathaniel Low referred to almanacs' importance in telling time when he boasted that they "serve as clocks and watches for nine-tenths of mankind." By the 1830s, however, almanacs were being replaced—at least in urban areas—by mass-produced clocks and watches, whose roles in bringing a semblance of order to commerce, transportation, and the workplace were appreciated by a society that was becoming increasingly market oriented.[20]

Demographic changes also affected the almanac trade. As the United States became more industrial and urban, the almanac, directed traditionally to a farming audience, became less relevant to an expanding middle-class readership. Many American printers and publishers began to specialize, concentrating on specific—and timely—topics and targeting specific audiences. The 1820s witnessed the emergence of a number of specialty almanacs. General almanacs soon began to face stiff competition from others concentrating on religion, politics, humor, agriculture, health and medicine, fraternal organizations, and the promotion of various antebellum reform movements.[21]

Astrological Rules

The introduction to *The Methodist Almanac* of 1846 outlined a history of almanacs up to that time. While the essay lauded almanacs for disseminating

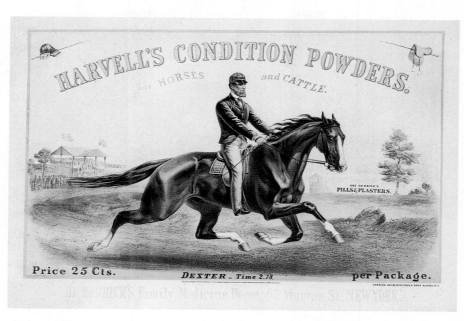

Plate 1. Harvell's Condition Powders for Horses and Cattle (Albany, N.Y.: Charles Van Benthuysen & Sons, 1869), chromolithograph. Courtesy of the author.

Plate 2. *If You Wish Perfect Health, Use the National Bitters,* Schlichter & Zug, Proprietors, Philadelphia (Philadelphia: R. F. Reynolds, 1867), chromolithograph. Courtesy of the Library Company of Philadelphia.

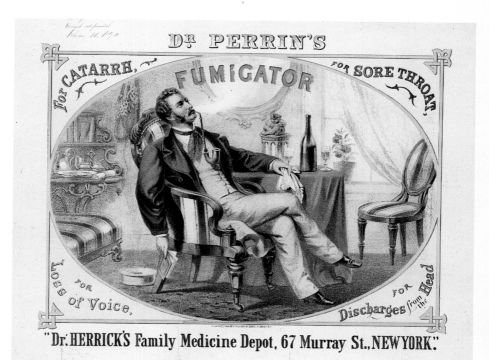

Plate 3. Dr. Perrin's Fumigator for Catarrh, for Sore Throat (Albany, N.Y.: Charles Van Benthuysen & Sons, 1869), chromolithograph. Courtesy of the New-York Historical Society.

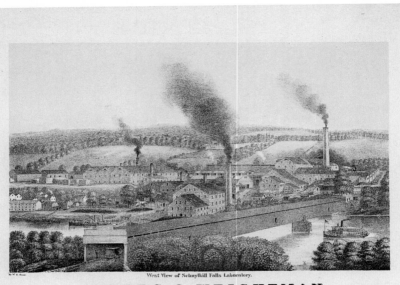

West View of Schuylkill Falls Laboratory.

POWERS & WEIGHTMAN
MANUFACTURING CHEMISTS,
PHILADELPHIA

OFFICE AT CITY LABORATORY
NINTH & PARRISH STS.

NEW YORK OFFICE
N° 56 MAIDEN LANE.

Plate 4. West View of Schuylkill Falls Laboratory, Powers & Weightman, Manufacturing Chemists, Philadelphia (Philadelphia: William H. Rease, ca. 1860), color lithograph. Courtesy of Philadelphia Museum of Art, The William H. Helfand Collection.

Plate 5. Indian Compound of Honey, Boneset and Squills (New York: Charles H. Hart, ca. 1870), chromolithograph. Courtesy of the New-York Historical Society.

Plate 6. Dr. Roback's Unrivaled Stomach Bitters, Prince, Walton & Co., Sole Proprietors, Cincinnati (Cincinnati: Gibson & Co., 1866), chromolithograph. Courtesy of the Library of Congress.

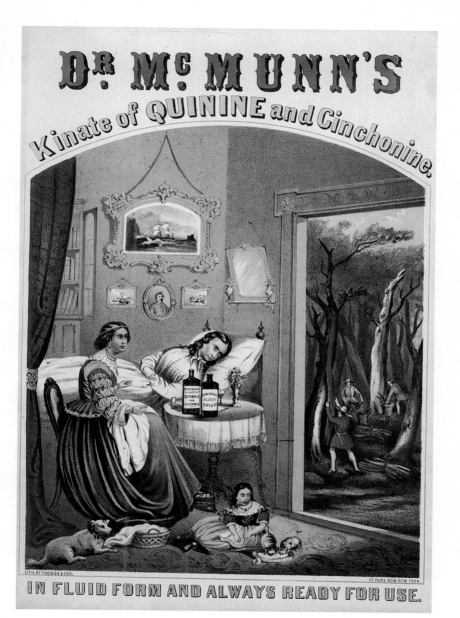

Plate 7. Dr. McMunn's Kinate of Quinine and Cinchonine, in Fluid Form and Always Ready for Use (New York: Thomas and Eno, ca. 1870), chromolithograph. Author's collection.

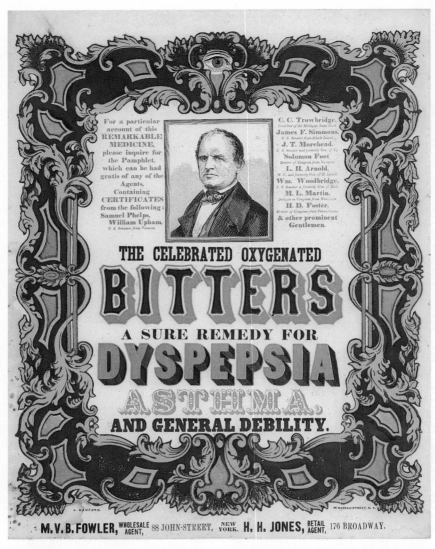

Plate 8. The Celebrated Oxygenated Bitters, a Sure Remedy for Dyspepsia, Asthma, and General Debility, M. V. B. Fowler, wholesale agent, H. H. Jones, retail agent (New York: A. Hanford, ca. 1846–47), color woodcut and relief print. Courtesy of Philadelphia Museum of Art, The William H. Helfand Collection.

guidance concerning life's everyday concerns, including health and sickness.[24] Its popularity was helped significantly by the appearance of various forms of print, particularly at the low end of the market, where almanacs and other cheap publications disseminated astrological advice as well as other health information to a wider population of readers.[25] Just how popular almanacs were during this period is apparent in the distribution figures, which show that during the mid–seventeenth century almanac sales averaged about four hundred thousand annually, which, according to Bernard Capp, "suggests that roughly one family in three bought an almanac."[26]

Astrological medicine was based on the belief that different signs of the zodiac ruled over specific parts of the body. Thus readers of almanacs were quite familiar with the image of the human body—usually a crude woodcut—popularly known as the "Zodiac Man," the "Man of Signs," or the "Anatomical Man." The Anatomical Man depicted the human body—sometimes a female body— surrounded by the twelve signs of the zodiac. This image was presented so the reader could see which parts of the body a particular sign governed as the moon passed through that sign. The moon's place was critical in terms of bloodletting because it was thought that the moon controlled the amount of blood in the veins. It was believed that when the moon was in a particular sign, the blood in the part governed by that sign would be at its fullest. Bleeding was proscribed at this time because it was thought that it would lead to a dangerous, even fatal, effusion of blood. For example, a physician or a practitioner would not bleed an individual from the chest area if the moon were to enter Leo.[27] Knowing the location of the moon in relation to the zodiac, the reader could then determine the appropriate times for not only letting blood but also administering medicines, gathering herbs, and performing surgical procedures. The Anatomical Man became a popular component of English almanacs as well as their American counterparts.[28]

Harvard almanacs for the most part shunned astrology because their compilers equated it with paganism. As a result, there were neither weather predictions nor cuts of the Anatomical Man in any of these publications. The first American almanac to include an image of the Anatomical Man was John Foster's 1678 annual. A graduate of Harvard, Foster had established the second printing press in Massachusetts (in Boston) and issued the first almanac to compete with those of the Harvard series. Foster did not make weather predictions, however.[29]

John Tulley's almanacs diverged more dramatically than did Foster's from

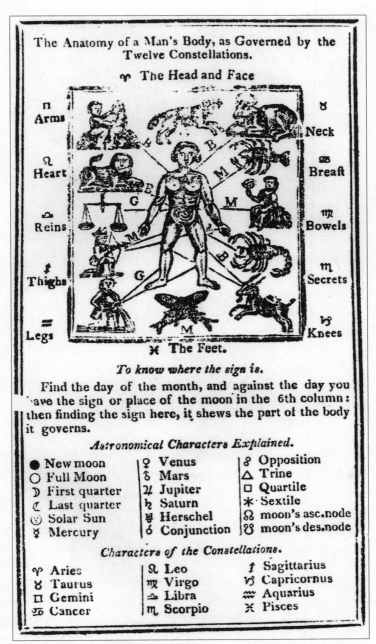

The Anatomy of a Man's Body, as Governed by the Twelve Constellations.

♈ **The Head and Face**

♊ **Arms**

♉ **Neck**

♌ **Heart**

♋ **Breaſt**

♎ **Reins**

♍ **Bowels**

♐ **Thighs**

♏ **Secrets**

♒ **Legs**

♑ **Knees**

♓ **The Feet.**

To know where the sign is.

Find the day of the month, and against the day you 'ave the sign or place of the moon in the 6th column: then finding the sign here, it ſhews the part of the body it governs.

Aſtronomical Characters Explained.

● New moon	♀ Venus	☍ Opposition
○ Full Moon	♂ Mars	△ Trine
☽ First quarter	♃ Jupiter	□ Quartile
☾ Last quarter	♄ Saturn	✳ Sextile
☉ Solar Sun	♅ Herschel	☊ moon's asc. node
☿ Mercury	☌ Conjunction	☋ moon's des. node

Characters of the Constellations.

♈ Aries	♌ Leo	♐ Sagittarius
♉ Taurus	♍ Virgo	♑ Capricornus
♊ Gemini	♎ Libra	♒ Aquarius
♋ Cancer	♏ Scorpio	♓ Pisces

"The Anatomy of a Man's Body," in *The North-American Calendar, or the Columbian Almanac* (Wilmington, Del.: Printed by Robert Porter, [1821]), [2]. Courtesy of the author.

The Anatomy of a Man's Body as Govern'd by the Twelve Constellations.

Half of the almanacs I consulted contained either the Anatomical Man or a list of the signs of the zodiac, while over 90 percent listed the moon's place on the monthly calendar pages.[37] Benjamin Franklin, for example, included the Anatomical Man in all but one of his twenty-five *Poor Richard* almanacs. He provided instructions in his first issue (1733) on how to link the diagram with the calendar: "First find the Day of the Month, and against the Day you have the Sign or Place of the Moon in the 5th Column. Then finding the Sign here, it shews [sic] the part of the Body it governs." Several almanacs carried the same directions; others did not, presumably because their compilers felt that it was unnecessary.[38] Occasionally, an almanac maker would resort to poetry to explain the Anatomical Man to readers. The 1768 edition of *The Virginia Almanack* offered these lines along with the image:

The Ram possesseth head and face,
The neck the Bull commands,
The loving Twins, with equal power,
Guide shoulders, arms, and hands.
The breast and stomach Cancer owns,
The heart the Lion claims,
The Virgin loves the belly piece,
To Libra loins and reins.
The Scorpion has the secret part,
The thighs to Sagittarius,
To Capricorn we give the knees,
The legs unto Aquarius.
Now none but Pisces wants a share,
To him we give the feet,
Then Aries head and face again,
And so make both ends meet.[39]

Some almanac makers, through editorial comments, expressed their faith in the astrological rules (natural astrology) they offered. The most ardent defender of astrology was Nathaniel Ames Sr., the most successful almanac maker of the eighteenth century. A physician and tavern keeper who was proficient in natural history, astronomy, and mathematics, Ames, in prefatory remarks to his 1764 almanac, asserted that astrology influenced human affairs and could, if used with "good Sense and Learning" and "lawful Means," lead men "on to Greatness": "[T]he cealestial [*sic*] Powers that can and do agitate and move the whole Ocean, have also Force and Ability to change and alter the Fluids and Solids of the humane [*sic*] Body; and that which can alter and change the Fluids and Solids of the Body, must also greatly affect the Mind: and that which can and does affect the Mind, has a great Share and Influence in the Actions of Men."[40]

Ames believed that planetary alignments and signs of the zodiac could influence an individual's health. And like many proponents of astrology, he was convinced that the moon played a critical role in influencing health and the weather.

Ames and other almanac makers who shared his views of astrology, however, were in the minority. Evidence suggests that the majority of almanac makers held at best skeptical and at worst harsh opinions of astrology. Some, like Samuel Clough, used poetry in his 1703 almanac to ridicule the Anatomical Man and readers who believed in it:

The Anatomy must still be in
Else the *Almanack's* not worth a pin;
For *Country-men* regard the Sign
As though 'Twere Oracle Divine
But do not mind that altogether,
Have some respect to Wind and Weather.[41]

Jacob Taylor, compiler of the *Pensilvania* almanac, used his 1746 issue to attack other almanac makers and readers who subscribed to astrology: "Is it not amazing Confidence, in Men destitute of Physical Learning, to persist so boldly in giving their Direction? They could never impose such Fooleries on the World, were not the Writers, and some weak Readers riveted to Superstition, and fastened with Nails of Idolatry."[42] The 1812 edition of *Johnson and Warner's Almanac* launched a scathing attack on astrologers, referring to them as "imposters who thrive by the folly of others." One cannot "foretel [*sic*] events many years before they happen," claimed the editor, "unassisted by revelation." The almanac urged its readers to banish from their minds the "vain" and "sinful curiosity" of predicting the future.[43] And *The Mechanics' Almanac, and Astronomical Ephemeris, for 1825* omitted the Anatomical Man because, its editors reasoned, "[I]n the present refined state of Science, it is high time that this last relic of Astrology . . . be expunged from our Almanacs; for however credulous

cations, notably those of Luigi Cornaro and George Cheyne, emphasized the six Galenic "nonnaturals": air, food and drink, sleeping and waking, exercise and rest, the evacuations, and the passions of the mind. Abuse of one or more of the nonnaturals could result in disease or even death; thus moderation was the key to a healthy and long life.[47] These regimen texts were to be read (presumably by affluent readers who had time for contemplation) but not actively applied, unlike the domestic medical guides, which belonged to the second traditional category of popular medical publications. Domestic medical texts, such as those of John Wesley and Nicholas Culpeper, contained, for the most part, descriptions of a variety of diseases and directions for treating them. Works in this category were most likely intended for literate artisans, farmers, and laborers—those who could not afford the services of a physician and for whom an extended illness could spell economic disaster. Directed to the widest possible audience of readers, early American almanacs (general almanacs) offered both categories of health advice.[48]

More than half of American general almanacs issued between the 1730s and 1860 contained medical "cures" or remedies for a wide variety of ailments, ranging from bruises, colds, corns, dropsy, earaches, headaches, rheumatism, and warts to intermittent fever (malaria), whooping cough, consumption, dysentery, and bladder stones.[49] While some of the almanac remedies appear to have been supplied by compilers or printers from other published sources (newspapers, domestic medical texts, or other almanacs), many were contributed by readers. Although the therapeutic advice in early American almanacs included a variety of chemical ingredients, it tended to emphasize plants, herbs, nuts, and barks that were readily available to readers.

The remedies in early American almanacs—like those presented in other lay health publications of the time—were based on shared assumptions about the body, health, and disease. These assumptions were profoundly influenced by the concept of humoralism. Formalized by Galen in the second century, this tradition interpreted health as a natural balance of the body's humors (yellow bile, blood, phlegm, and black bile) and their respective qualities (hot and dry, hot and moist, cold and moist, cold and dry). The body was viewed in holistic terms, as an equilibrium system that constantly interacted with its environment. This interaction was visible in the body's system of intake and outgo, a process through which the body constantly strove to achieve a harmonious balance of its humors. Illness was the result not of a specific disease but of imbalance, and prolonged imbalance could lead to death. Health was restored when balance was restored. The primary role of therapeutics, then, was to restore this balance by either depleting the system by a variety of measures, such as bleeding; administering cathartics (for inducing defecation), diuretics (for inducing urination), and emetics (for inducing vomiting); or strengthening the system through the use of astringents and tonics. The remedies encountered in early American almanacs were composed of plant, vegetable, and mineral substances that produced a predicted physiological effect, such as sweating, salivation, or evacuation. Both the almanac maker and the almanac reader assumed that, in an age when trained physicians were few and expensive, the maintenance and the restoration of health were the layperson's responsibility.[50]

Although the humoral tradition remained dominant in lay practice well into the nineteenth century, it was not the only system that influenced lay therapeutic practices. Lay as well as professional therapeutics incorporated elements of the Paracelsian tradition into their humoral approach. In contrast to the Galenic system, Paracelsus (1482–1546) and his followers asserted that disease was not simply an imbalance of humors but a specific entity that had a specific external cause. And because diseases were specific entities, they required specific remedies. Paracelsus and his followers recommended various mineral or chemical remedies, such as mercury, antimony, and arsenic, for specific disorders. Although the Paracelsian challenge failed to replace traditional humoralism in American lay and professional practice, its influence is readily apparent in the various chemical ingredients that appeared in early American pharmacopoeias, professional medical texts, domestic medical guides, family recipe books, and, of course, almanacs.[51]

Remedies for dropsy and dysentery that appeared in early American almanacs provide useful examples of the therapeutic assumptions discussed above. "Dropsy" was the term used in the eighteenth and early nineteenth centuries to describe an abnormal accumulation of fluid in the body. Thus a reader would not have been surprised—and would, in fact, have expected—to encounter a remedy such as the one printed in the 1787 *Wilmington Almanack* that would result in depletion, or in this case, urination. The *Wilmington Almanack*'s "Cure for Dropsy" was submitted by one Jonathan Zane, presumably a reader of the publication. Zane, at one time much "swelled," was unable to cure himself of the disorder. After failing to locate a trained physician who could relieve his distress, Zane eventually stumbled upon a remedy and wanted to share it with other readers. After consuming an "ounce of Salt Petre dissolved in a pint of cold water" every morning and evening for five days, his "evacuations of water were increased," and he was "much relieved."[52] More than three decades later, the *Farmer's Almanac* (New York) ran a cure for dropsy that, although comprising different ingredients, produced the same result. The almanac's "recipe" called for "a double handful of parsley roots . . . a handful of horse radish scraped, two table spoonsful [*sic*] of pounded mustard seed," and "half an ounce of juniper berries." "After the water" had "passed off," the patient was advised to regain strength by engaging in "moderate exercise," subsisting on "nourishing food," and abstaining from liquor "as much as possible."[53]

The dropsy remedies induced evacuation in order to reduce an unnatural accumulation of fluid in the body. This makes perfect sense when considered within the context of humoralism. A twenty-first-century observer looking back at early American almanac remedies for dysentery might think that, in keeping with the concept of humoral balance, they would consist of ingredients that would strengthen or build up a system depleted or weakened by the loss of fluids.[54] This was not the case, however. In fact, almanac remedies for this complaint included bleeding and blistering while promoting various forms of evacuation. For example, the 1797 issue of *Poulson's Town and Country Almanac* offered its readers "Useful hints for persons in the country concerning the Bloody Flux" that, for the most part, resulted in depletion. The "hints" presented by this Philadelphia almanac consisted of two remedies, and the presence or absence of fever determined which one was appropriate:

> It sometimes happens that the *Dysentery* or *Bloody Flux*, occurs without much fever. This kind may soon be cured, by taking a table-spoonful of Caster-oil, or an ounce of Glauber's salts every morning, and thirty or forty drops of liquid Landanum [Ladanum] every night, for four or five times successively; but it is most common for this disease to be attended with fever; in this case, besides the above mentioned treatment, it will be necessary to bleed two or three times, and if the disorder does not quickly yield, ten or twelve grains of Jalap or Rhubarb, and six or eight grains of Calomel, instead of the oil or salts every day. Blisters must be applied to the wrists and ancles [*sic*]: Rennet-whey is the most proper drink—Heating remedies must be avoided.[55]

Many physicians and lay practitioners of the time believed that fever was present in most cases of dysentery. This may explain why almanac remedies for the complaint included ingredients that would reduce the fever as well as restore natural bodily evacuations disrupted by the fever. *Smith & Forman's New-*

York & New-Jersey Almanac for 1816 provides such an example. This issue included a "Cure for the Dysentery" that consisted of "1 ounce of rheubarb [*sic*], 2 drachms of English saffron," and "2 or 1½ drachms of cardimen [cardamom] seed" mixed with "one pint of good French Brandy."[56] That same year *The North-Carolina Almanack* offered its readers "A Receipt for the Flux" that produced similar results: "Take of Emetic Tartar 10 grains. Epicacuana [ipecacuanha] 20 grains. Jolop [jalap] 40 grains make them into 30 pills with honey & flour, the dose one pill three times a day. The diet should be chiefly of mutton & Chicken broth. When the flux attacks violently, a blister aplied [*sic*] over the pain is often useful and in such cases the use of the above pills should be preceeded [*sic*] by a gentle purge."[57]

A twenty-first-century observer might well ask why anyone would prescribe a treatment consisting of ingredients that promoted evacuations of one sort or another in a patient already weakened by the bloody flux. While these dysentery remedies seem to contradict the basic assumptions of humoralism, they were, in fact, very much in keeping with the tenets of the theory as they were understood by professional and lay practitioners of the period. Although there may have been a lack of consensus among these practitioners concerning the cause, appearance, and nature of fever, most of them were advocating treatments for the disorder that tended to be, for the most part, a mix of depletive and stimulating measures. Because fever was generally thought to obstruct the natural bodily evacuations, thus disrupting the body's balance or equilibrium, many professional and lay practitioners believed its presence during a complaint like dysentery must be addressed when commencing treatment. Simply put, dysentery remedies that appeared in early American almanacs—as well as in scores of domestic health guides and professional textbooks—were intended to subdue fever, remove inflammation from the intestines, decrease pain associated with the complaint, and restore normal evacuations by promoting defecation, urination, and sweating.[58]

Recall that up through the 1750s American almanacs occasionally offered readers advice concerning the astrological properties of herbs and plants, with particular emphasis on the appropriate times for gathering them. Few almanacs, however, actually described the herbs, plants, and minerals that composed the cures they published. Presumably, almanac makers and publishers assumed that their readers were familiar with the botanical and mineral components of the remedies, that readers understood the properties of each component, and that readers knew what physiological results to expect.

Remedies were for restoring health; regimens were for preserving it by espousing a life of moderation—or temperance, as it was then broadly defined—
in all things involving the nonnaturals. Moderation involving the nonnaturals
was in keeping with the humoral assumptions concerning the body, health, and
disease. Isaiah Thomas, for example, included in his 1781 almanac an extract
from "an excellent" but unnamed author on the "Art of Preserving Health"; the
extract addresses the importance of moderation in preserving the body's delicate balance: "The whole art of preserving health may be said to consist of fill-
ing up what is deficient, and emptying what is redundant, that the body may
be kept in its natural state; and therefore all the supplies from eating and drinking, and all the discharges by sweating and other channels of nature, should be
regulated, that the body may not be oppressed by repletion, nor wasted by
evacuation."[59]

The connection between long life and personal behavior was part of a traditional medical worldview that dated back to antiquity and was later popu-
larized by scores of texts promoting healthy living. Early American almanacs
were filled with warnings about the dangers to health posed by such vices as
drunkenness, gluttony, indolence, and uncontrolled passions. That drunkards,
gluttons, the indolent, and the morally deficient, through their behavior, predispose themselves to an array of diseases is a theme almanac readers encountered frequently.

Regimen advice in almanacs was usually conveyed through essays, maxims,
and poetry. Nathaniel Ames Sr., for example, included the maxim "Live temperate and defy the Physician" in his 1761 almanac. His son, Nathaniel Ames
Jr., used poetry in his 1767 annual to advance a temperance message:

'Tis to thy rules, O Temperance! we owe
All pleasures, which from health and strength can flow,
Vigour of body—purity of mind,
Unclouded reason—sentiments refin'd,
Unmix'd, untainted joys—without remorse,
Th' intemperate Sensualist's never-failing curse.[60]

The compiler of *The Columbian Calendar, or New-York and Vermont Almanack* for 1817 echoed Ames's view with a verse of his own:

In vain we mourn those transitory days
Consum'd in riot and licentious ways;

'Tis Temperance alone preserves our strength,

In mind and body to life's utmost length.[61]

Essays on the specific dangers of too much or too little sleep, too much or too little exercise, indolence, and tobacco appeared frequently. The evils of drinking and overeating, however, received the most attention. Essay after essay, poem after poem, and maxim after maxim condemned drunkenness and gluttony. In a brief essay on "The Drunkard's Looking Glass," Isaiah Thomas presented readers of his 1799 almanac with a graphic description of alcohol's deleterious impact on the human body: "Gout-sickness-puking-tremors of the hands in the morning, bloatedness, inflamed eyes-red nose and face-sore and swelled legs-jaundice-pains in the limbs, and burning in the hands and feet-dropsy-epilepsy-melancholy-idiotism-madness-palsy-apoplexy-death."[62]

Losing control of one's passions was deemed just as dangerous as fouling one's stomach with rich food and strong drink. The 1855 issue of *Phinney's Calendar or Western Almanac* instructed its readers about the important link between a healthy mind and a healthy body: "[A]cquire a composure of mind and body. Avoid agitation or hurry of one or the other, especially just before and after meals. To this end govern your temper—endeavor to look at the bright side of things—discard envy, hatred and malice, and lay your head upon your pillow in charity with all mankind." Three years earlier, *Middlebrook's New-England Almanac* had conveyed a similar message when it warned its readers about the harmful effects of novel reading: "A physician in Massachusetts says, 'I have seen a young lady with her table loaded with volumes of fictitious trash, poring, day after day and night after night, over highly wrought scenes and skillfully portrayed pictures of romance, until her cheeks grew pale, her eyes became wild and restless, and her mind wandered and was lost—the light of intelligence passed behind a cloud and her soul was for ever benighted. She was insane, incurably insane from reading novels.'"[63]

Health Advice and Specialization in the Almanac Trade

The pre–Civil War shift from general to specialty almanacs influenced the health advice offered to antebellum almanac readers. From the end of the War of 1812 to the Civil War, the number of specialty almanacs increased significantly. Several of these publications carried no health advice of any kind. Of those almanacs that did contain health advice, their compilers fashioned that advice to fit a particular agenda. Moreover, health advice found in specialized almanacs tended to espouse prevention more than treatment.

Porter's Health Almanac (Philadelphia, 1832–33), *Thomson's Almanac* (Boston, 1840–44), and *The Health Almanac* (New York, 1842–44) are three examples of specialty almanacs that offered health advice that was similar to and yet somewhat different from that traditionally found in general almanacs. The first of these was issued by Henry H. Porter of Philadelphia, who sought to establish himself in the book trade by publishing works on personal health and hygiene and on general self-improvement.[64] Porter's most successful venture was the *Journal of Health* (Philadelphia, 1829–33), which was one of this country's first medical periodicals published for the layperson.[65] In the autumn of 1831, Porter, in an attempt to reach an even wider audience, published the first issue of *Porter's Health Almanac*. Co-edited by Philadelphia physicians John Bell and D. Francis Condie, editors of Porter's *Journal of Health*, Porter's almanac offered the same advice as that publication: the value of a proper diet, regular exercise, cleanliness, and control of the passions.

Linking the traditional belief in the importance of personal hygiene to long life with the reform movements of the period, Bell and Condie espoused moderation and morality as the key elements of perfect health. Considering the professional interests of his editors, it is not surprising that Porter's almanac included neither astrological health advice nor home remedies. As members of a profession under attack by competing medical sects, Bell and Condie concerned themselves with establishing the authority of the medical profession in the management of disease. By circumscribing the role of the laity in medicine to prevention only, the co-editors attempted to strengthen the position of the trained physician in the sphere of treatment.[66] Yet the doctors' message of moderation and morality should not be seen merely as an attempt to promote their professional interests. Their message was also shaped by their attitudes concerning the body, health, and disease. The editors' regimen implied a conception of the body as an equilibrium system governed by—and embodying—the laws of nature created by God. Thus, living in accordance with God's laws would keep one healthy or restore one's health. Disease was the result of unnatural behavior, such as overeating, drinking too much, keeping late hours, or otherwise transgressing against the laws of nature.[67]

Samuel Thomson (1769–1843), whose fierce opposition to the therapeutic practices of the regular medical profession ignited a social movement that achieved national prominence during the antebellum period, published an almanac in Boston during the last years of his life. Thomson used his almanac to castigate the medical profession while espousing self-treatment based on his own medical philosophy. Thomson, too, saw the body as a delicate equilibrium system of intake and outgo. Putting his own spin on the traditional humoral theory, Thomson asserted that the body was healthy when its four components—earth, water, air, and fire (heat)—were balanced. Disease, on the other hand, was the result of imbalance, which, according to Thomson, diminished the body's natural heat. The purpose of Thomson's six-step therapeutic regimen, which used emetics, purgatives, steam or vapor baths, and herbal medicines, was to cure disease by restoring the body's natural heat.[68]

Thomson preached about the health benefits that would result from cleanliness, appropriate dress, exercise, the avoidance of alcohol and tobacco, and a proper diet. *Thomson's Almanac* advised its readers to avoid "regular doctors" and their "depletive and poisoning system" of bloodletting and mineral drugs.[69] In his 1843 almanac, Thomson advanced a simple rule for maintaining and restoring health: "Rise early, eat moderately, live an active and useful life, cultivate all the virtues and avoid the vices of society, and as soon as unwell use the Thomsonian means for cure."[70] He provided his readers with the information they needed to use his remedies, including descriptions (with woodcut illustrations) of various herbs and plants. Of his four Boston almanacs, only two contained astrological advice, and that was confined to the column in the monthly calendar that designated the moon's place.

The Health Almanac, published by John Burdell, a New York dentist, promulgated notions of the body, health, and disease similar to those found in *Porter's Health Almanac* and *Thomson's Almanac*. Burdell, an active participant in the health reform movement of the 1840s, advocated moderation and moral living as the keys to achieving and maintaining health.[71] His almanac went much further than either Porter's or Thomson's in its presentation of the body, however. Burdell provided his readers with detailed descriptions of the various organs and systems of the body—particularly those associated with the digestive tract. These descriptions were often accompanied by meticulous woodcut illustrations that depicted the body's intricate design. In this, Burdell was mirroring the trend of teaching anatomy and physiology in American common schools (tax-supported, state-regulated public schools that began to appear in the 1830s) as well as taking advantage of advances in printing technology.[72]

In terms of diet, Burdell advocated vegetarianism. Referring to himself as "A Vegetable Eater," Burdell argued that man was intended by God to subsist "upon fruit and vegetables, the natural productions of the earth." Man courts debility and disease when he "departs from this great first principle" and "violates the law of nature."[73] Burdell also promoted the teaching of anatomy and physiology in common schools as well as the reform of dress. Burdell did include the Anatomical Man and a column in each month's calendar for the moon's place and parts of the body affected. He did not offer his readers therapeutic advice, however, despite the fact that he was, like Samuel Thomson, highly critical of the regular medical profession. Burdell condemned regular physicians for what he believed was their growing use of mineral drugs. These types of drugs, according to Burdell, played no role in the restoration of health, but only exacerbated disease. "We need not expect to escape the penalty of violated physical law," Burdell asserted, "by taking medicine." By resorting to drugs, he warned, "we involve ourselves in deeper transgressions, and thereby incur a heavier punishment in the end."[74] Readers of Burdell's *Health Almanac* learned that health was achieved by living in harmony with the laws of nature, and that disease was an unnatural state caused by flouting these laws.

Almanacs and Popular Health in America

The productions of the early American almanac trade, whether general almanacs geared to a specific region of the country or specialty almanacs promoting a specific social movement, were, like the Bible, honored guests at many a fireside. Americans—the highly educated as well as the barely literate, the affluent as well as the poor, the urban dweller as well as the rural resident—sought not only almanacs' companionship but also their advice on many issues, including health. By the 1850s the golden age of almanacs was ending, however. A genre that once dominated the market of secular print was now competing with scores of newspapers, magazines, and books. By the middle of the century American almanac makers and publishers were trying to hold on to a tenuous niche in an increasingly competitive market. Leading cultural mediators in the eighteenth century, almanac makers were forced to become aggressive entrepreneurs in the nineteenth.

Because few astrology texts of any kind were published in America prior to the middle of the nineteenth century, almanacs played a leading role in disseminating astrology and astrological health information in early America.[75] Almanac makers correctly sensed that while astrology was scorned by the educated and the worldly, it remained popular with a significant segment of almanac readers, many of whom still clung to the rituals and practices of folk-lore and folk medicine well into the nineteenth century. Despite their own views of astrology, almanac makers knew that to succeed in the competitive market of print they could ill afford to alienate this important constituency. On the other hand, they strove to minimize the amount of astrology they presented to their readers and, as we have seen, openly criticized it. By minimizing as-trology and including material in their publications that would appeal to a more sophisticated audience, almanac makers hoped to retain those readers who looked upon astrology as a relic from an uncivilized past or as a mere joke. This was a risky endeavor, but it seemed to work—at least until the Civil War. Generally, almanac makers over time changed the way they presented astrol-ogy to their readers—from an active, instructional approach to a more am-bivalent, passive approach. However they presented astrology, almanac mak-ers were not only responding to the desires of a key segment of their readership but helping astrology survive (if only within certain segments of American so-ciety) and shaping contemporary and future perceptions of the practice.

In terms of treatment and regimen advice, almanacs played a complemen-tary role. The almanac was one of several genres of print through which lay health advice was disseminated in early America. One seeking a cure for a par-ticular ailment or advice on how to live a healthy life could consult numerous domestic medical guides, regimen texts, popular health periodicals, newspa-pers, domestic economy books, and publications of various sectarian groups. An almanac's "cures" may have been collected in a family recipe book or com-monplace book, or tipped into a domestic medical guide. An almanac's advice concerning healthy living may have been memorized or recorded in a journal for later use. Since early American print culture offered a diverse array of sources of health information for the layperson, the almanac may not have been the first place a person would have looked for health advice. Rather, the "cures" and health advice contained in almanacs—and encountered else-where—would have both reinforced and shaped readers' assumptions con-cerning the body, health, and disease.

Finally, did almanac readers actually follow the health advice offered by early American almanacs (or, for that matter, other popular health publica-tions)? Were lay Americans bleeding and purging themselves, as implied by the astrological rules and the remedy advice offered by almanacs? The lack of solid evidence prevents one from offering a definitive answer.[76] Yet similar advice pervaded other forms of popular health literature as well as commonplace books, journals, and family recipe books. Moreover, popular health publications, like almanacs, were issued in numerous editions during the first half of the nineteenth century, which implies that there was a demand for these works among American consumers. Finally, one must consider the fact that many almanac "cures" were contributed by readers. Thus one can infer that at least some readers followed almanac health advice and believed in the assumptions upon which it was based. In closing, I would argue that the rules, remedies, and regimens offered by early American almanacs both reflected and helped shape popular assumptions concerning the body, health, and disease.

NOTES

1. Marion Barber Stowell, "The Influence of Nathaniel Ames on the Literary Taste of His Time," *Early American Literature* 18 (1983): 128; Samuel Briggs, *The Essays, Humor, and Poems of Nathaniel Ames, Father and Son, of Dedham, Massachusetts, From Their Almanacs 1726–1775* (Cleveland: The Subscribers, 1891), 13. Moses Coit Tyler, in his 1878 study of American literature, summed up current opinion of almanacs when he referred to the genre as "the very quack, clown, pack-horse, and pariah of modern literature." *A History of American Literature*, 2 vols. (New York: G. Putnam's Sons, 1878), 1:120.

2. Briggs, *Essays*, 13.

3. Clarence S. Brigham, one-time librarian of the American Antiquarian Society, was an early advocate of using almanacs to study American popular culture. See his "Report of the Librarian," *Proceedings of the American Antiquarian Society* 35 (1925): 190–218.

4. Examples of how almanacs can be used to study American popular culture are David D. Hall, *Worlds of Wonder, Days of Judgment: Popular Religious Belief in Early New England* (Cambridge: Harvard University Press, 1989); Stephen Nissenbaum, *The Battle for Christmas* (New York: Alfred A. Knopf, 1997); Elizabeth Carroll Reilly, "Common and Learned Readers: Shared and Separate Spheres in Mid-Eighteenth-Century New England" (Ph.D. diss., Boston University, 1994); and David H. McCarter, "'Of Physick and Astronomy': Almanacs and Popular Medicine in Massachusetts, 1700–1764" (Ph.D. diss., University of Iowa, 2000).

5. This article is based primarily on the examination of almost nine hundred almanacs (both general and specialized) published in America between 1647 and 1860. Most of the almanacs I consulted, however, were issued between 1750 and 1860. Of the almanacs examined, 92 percent contained some form of astrological health advice, 64 percent included remedies for various ailments, and 21 percent offered regimen advice. On the face of it, the percentages regarding astrological advice, remedies, and regimen advice are misleading, for several specialized almanacs that contain no health advice tend to skew these percentages. The percentages for astrological advice and remedies are higher in general almanacs. On the other hand, regimen advice appears more frequently (47 percent) in specialized almanacs. Early American almanacs contained other forms of health and medical information, such as descriptions of various herbs and plants and

explanations of how to use them, essays on recent epidemics, extracts from newspapers and professional medical publications, maxims, and patent medicine advertisements and testimonials.

6. Milton Drake, comp., *Almanacs of the United States*, 2 vols. (New York: Scarecrow Press, 1962), 1:ii; Barnard Capp, *English Almanacs, 1500–1800: Astrology and the Popular Press* (Ithaca: Cornell University Press, 1979). See also George Parker Winship, *The Cambridge Press, 1638–1692* (Philadelphia: University of Pennsylvania Press, 1945).

7. Capp, *English Almanacs*.

8. Many eighteenth- and nineteenth-century almanac makers hired "philomaths," or local mathematicians and/or astronomers, to provide astronomical and astrological calculations (including weather predictions) for their publications.

9. Marion Barber Stowell, *Early American Almanacs: The Colonial Weekday Bible* (New York: Burt Franklin, 1977). See also John T. Kelly, *Practical Astronomy during the Seventeenth Century: Almanac-Makers in America and England* (New York: Garland Publishing, 1991); Robb Sagendorph, *America and Her Almanacs: Wit, Wisdom, and Weather, 1639–1970* (Boston: Little, Brown and Company, 1970); and Drake, *Almanacs of the United States*. By 1801 American printers had issued twenty-four hundred different almanacs. McCarter, "Of Physick and Astronomy," 13.

10. The term *aspect* refers to the angular relationship between various points or planets in the horoscope (an astrological chart). The five major aspects listed in almanacs were conjunction (0 degrees), sextile (60 degrees), quartile or square (90 degrees), trine (120 degrees), and opposition (180 degrees). James R. Lewis, *The Astrology Encyclopedia* (Detroit: Visible Ink Press, 1994), 40–43.

11. The zodiac is the "belt" constituted by the twelve signs—Aries, Taurus, Gemini, Cancer, Leo, Virgo, Libra, Scorpio, Sagittarius, Capricorn, Aquarius, and Pisces. The foundation of medical astrology is a system of correspondence between the twelve signs and various parts of the human body. The Anatomical Man was a pictorial representation of that relationship. Ibid., 354–57, 535–37. Few almanacs offered both the list of signs of the zodiac and the Anatomical Man; most offered either one or the other.

12. This information sometimes included "Vulgar Notes," which indicated numbers, letters, and days thought to have magical, superhuman significance for the coming year. John Butler, "Magic, Astrology, and the Early American Religious Heritage, 1600–1760," *American Historical Review* 84 (1979): 330.

13. Unlike most forms of print, almanacs posed minor risks for colonial American printers. Because they had difficulty in obtaining basic manufacturing materials, colonial printers could ill afford the financial risk of producing full-length books for a market that was virtually nonexistent. Books, be they original American works or reprints of English and European publications, were expensive to produce, and thus their retail price was beyond the financial reach of most American readers. As a result, most printers made a living selling such merchandise as stationery and patent medicines, offering their services as a book binder and a postmaster, and printing cheap publications such as blank forms, single-sheet newspapers, broadsides, and almanacs. John Bidwell, "Printers' Supplies and Capitalization," in *A History of the Book in America*, vol. 1, *The Colonial Book in the Atlantic World*, ed. Hugh Amory and David D. Hall (Cambridge: Cambridge University Press; Worchester, Mass.: American Antiquarian Society, 2000), 163–64; James N. Green, "English Books and Printing in the Age of Franklin," in ibid., 266; Hugh Amory, "Reinventing the Colonial Book," in ibid., 45, 51. See also Lawrence

C. Wroth, *The Colonial Printer* (Charlottesville: University Press of Virginia, 1964); Hell-mut Lehmann-Haupt, *The Book in America: A History of the Making and Selling of Books in the United States* (New York: R. R. Bowker, 1952); Rollo Silver, *The American Printer, 1787–1825* (Charlottesville: University Press of Virginia, 1967); and John Tebbel, *A History of Book Publishing in the United States*, vol. 1, *The Creation of an Industry, 1630–1865* (New York: R. R. Bowker, 1972).

14. Wroth, *Colonial Printer*, 228; Hall, *Worlds of Wonder*, 58. Almanacs were sold in bookshops, in general stores, and by printers, merchants, and peddlers. They also circulated in taverns, coffeehouses, and post offices. Stowell, *Early American Almanacs*, 30–31.

15. Benjamin Franklin, *Writings* (New York: Library of America, 1987), 1397, quoted in Douglas Anderson, *The Radical Enlightenments of Benjamin Franklin* (Baltimore: Johns Hopkins University Press, 1997), 106; Low, *An Astronomical Diary, or Almanack for . . . 1762* (Boston: D. & J. Kneeland, [1761]), quoted in Reilly, "Common and Learned Readers," 246; and Ames, *An Astronomical Diary: or an Almanack, for the Year . . . 1754* (Boston: Printed by J. Draper, [1753]), [1], quoted in Briggs, *Essays*, 249.

16. For the classical influence in early American almanacs, see Richard M. Gummere, "The Classical Element in Early New England Almanacs," *Harvard Library Bulletin* 9 (1955): 181–96. *The Farmer's Almanack, for the Year . . . 1811* (Baltimore: Printed and Sold by Warner & Hanna, [1810]) contained "Directions for Servants," indicating that its compiler envisioned a readership that would have included individuals and families affluent enough to afford domestic and/or enslaved help. Almanacs were often interleaved with blank sheets of paper and used as diaries and account books. Reilly studied several almanac-diaries and found that they were "used by urban as well as rural people and by wealthy and common readers." Reilly, "Common and Learned Readers," 407.

17. For a discussion of almanac makers particularly and printers generally as cultural mediators, see Reilly, "Common and Learned Readers," 330, 409–11. The social background of almanac makers and printers is discussed in Sara S. Gronim, "At the Sign of Newton's Head: Astronomy and Cosmology in British Colonial New York," *Pennsylvania History* 66 (1999): 58; and Hugh Amory, "The New England Book Trade, 1713–1790," in *Colonial Book*, ed. Amory and Hall, 1:334. Amory cites the Federal Census of 1790, which placed Boston printer-booksellers at the top of the artisan class, just below sea captains.

18. A survey of Drake's *Almanacs of the United States* demonstrates the rate at which almanacs proliferated during the eighteenth century and the first half of the nineteenth. According to Drake, there were in 1700 only 4 almanac titles being printed in America. This number had expanded to 25 titles by 1750 and had reached 135 titles by 1800. The number of almanacs increased at an even higher rate during the first half of the nineteenth century.

19. The expansion of popular print in late-eighteenth- and early-nineteenth-century America has received increased scholarly attention over the last fifteen years or so. See, for example, Victor Neuberg, "Chapbooks in America: Reconstructing the Popular Reading of Early America," in *Reading in America: Literature and Social History*, ed. Cathy N. Davidson (Baltimore: Johns Hopkins University Press, 1989), 81–113; David Paul Nord, "A Republican Literature: Magazine Reading and Readers in Late-Eighteenth-Century New York," in ibid., 114–39; Daniel A. Cohen, *Pillars of Salt, Monuments of Grace: New England Crime Literature and the Origins of American Popular Culture* (New York: Oxford

University Press, 1993); David S. Reynolds, *Beneath the American Renaissance: The Subversive Imagination in the Age of Emerson and Melville* (Cambridge: Harvard University Press, 1988); Reilly, "Common and Learned Readers"; and Isabelle Lehuu, *Carnival on the Page: Popular Print Media in Antebellum America* (Chapel Hill, N.C.: University of North Carolina Press, 2000). For the rise of newspapers and magazines in America, see Charles E. Clark, *The Public Prints: The Newspaper in Anglo-American Culture, 1664–1740* (New York: Oxford University Press, 1994); Michael Schudson, *Discovering the News: A Social History of American Newspapers* (New York: Basic Books, 1978); and the introduction to Kenneth M. Price and Susan Belasco Smith, eds., *Periodical Literature in Nineteenth-Century America* (Charlottesville: University Press of Virginia, 1995).

20. Low, *An Astronomical Diary: or Almanack for . . . 1786* (Boston: T. & J. Fleet, [1785]), quoted in Drake, *Almanacs of the United States*, 1:viii. See Michael O'Malley, *Keeping Watch: A History of American Time* (New York: Viking, 1990); and Ian R. Bartky, *Selling True Time: Nineteenth-Century Timekeeping in America* (Stanford, Calif.: Stanford University Press, 2000) for two concise studies of the advent of the clock and wristwatch in America. Charles Sellers's *The Market Revolution: Jacksonian America, 1815–1846* (New York: Oxford University Press, 1991) provides a useful description of how the rapid development of the marketplace transformed American life in the decades following the War of 1812.

21. Drake, *Almanacs of the United States*, 1:xii; Michael A. Lofaro, *The Tall Tales of Davy Crockett: The Second Nashville Series of Crockett Almanacs, 1839–1841* (Knoxville: University of Tennessee Press, 1987), xvi–xix; Tebbel, *History of Book Publishing*, 1:545–46.

22. *The Methodist Almanac for the Year . . . 1846* (New York: Published by Lane & Tippett for the Methodist Episcopal Church, [1845]), [3].

23. Carole Rawcliffe, *Medicine and Society in Later Medieval England* (Stroud, England: Alan Sutton Publishing, 1995), 83–85. See also Peter Whitfield, *Astrology: A History* (New York: Harry N. Abrams, 2001); S. J. Tester, *A History of Western Astrology* (Woodbridge, England: Boydell Press, 1987); and Patrick Curry, *Prophecy and Power: Astrology in Early Modern England* (Oxford: Polity Press, 1989). Throughout the Middle Ages and well into the modern period, there was general agreement among educated people that the planets and stars exerted a profound influence over the earth and the weather. Peter Eisenstadt, "The Weather and Weather Forecasting in Colonial America" (Ph.D. diss., New York University, 1990), 60.

24. Capp, *English Almanacs*, 15–16, 20–21, 64, 180, 190–91; Keith Thomas, *Religion and the Decline of Magic* (New York: Charles Scribner's Sons, 1971), 283–85, 296. See also Michael McDonald, "The Career of Astrological Medicine in England," in *Religio Medici: Medicine and Religion in Seventeenth-Century England*, ed. Ole Peter Grell and Andrew Cunningham (Brookfield, Vt.: Scolar Press, 1996), 62–90; idem, *Mystical Bedlam: Madness, Anxiety, and Healing in Sixteenth-Century England* (Cambridge: Cambridge University Press, 1981); Barbara H. Traister, "Medicine and Astrology in Elizabethan England: The Case of Simon Forman," *Transactions and Studies of the College of Physicians of Philadelphia*, ser. 5, 11 (1989): 279–97; idem, *The Notorious Astrological Physician of London: Works and Days of Simon Forman* (Chicago: University of Chicago Press, 2001); Allan Chapman, "Astrological Medicine," in *Health, Medicine, and Mortality in the Seventeenth Century*, ed. Charles Webster (Cambridge: Cambridge University Press, 1979), 275–300; and Anna Marie Roos, "Luminaries in Medicine: Richard Mead, James Gibbs,

and Solar and Lunar Effects on the Human Body in Early Modern England," *Bulletin of the History of Medicine* 74 (2000): 433–57.

25. Thomas, *Religion and the Decline of Magic*, 288–89. Elizabeth Eisenstein's *The Printing Press as an Agent of Change* (Cambridge: Cambridge University Press, 1979) provides a useful overview of print culture and the popularization of print in the late fifteenth century and the early modern period. Margaret Spufford's *Small Books and Pleasant Histories: Popular Fiction and Its Readership in Seventeenth-Century England* (Athens: University of Georgia Press, 1982) provides a useful overview of the rise of the English market in popular literature. Popular health publications, such as the various editions of *Erra Pater*, were serious competitors of the almanac at the low end of the print market in England. For an excellent introduction to *Erra Pater*, see Mary Fissell, "Readers, Texts, and Contexts: Vernacular Medical Works in Early Modern England," in *The Popularization of Medicine, 1650–1850*, ed. Roy Porter (London: Routledge, 1992), 72–96.

26. Capp, *English Almanacs*, 23. Capp cites distribution figures presented in Cyprian Blagden, "The Distribution of Almanacs in the Second Half of the Seventeenth Century," *Studies in Bibliography* 11 (1958): 108–17, an article that I also have consulted.

27. Rawcliffe, *Medicine and Society*, 86–87; Thomas Palmer, *The Admirable Secrets of Physick and Chryrurgery*, ed. Thomas Rogers Forbes (New Haven: Yale University Press, 1984), 3. A scholarly study of the history of medical astrology has yet to be done. Charles Arthur Mercier's *Astrology in Medicine* (London: Macmillan and Co., 1914), though dated, is still useful. Reinhold Ebertin's *Astrological Healing: The History and Practice of Astromedicine* (York Beach, Maine: Samuel Weiser, 1989) concentrates on the late nineteenth and twentieth centuries.

28. Capp, *English Almanacs*, 24, 191; Thomas, *Religion and the Decline of Magic*, 287, 324; Louis Winkler, "Technical Aspects of Eighteenth-Century Common Almanacs," *East-Central Intelligencer* 6, no. 3, n.s. (1992): 11–12; and Chapman, "Astrological Medicine," 293–94.

29. Eisenstadt, "Weather and Weather Forecasting," 93–94.

30. Ibid., 93–95, 99; *1695. The New England Almanack for the Year . . . MDCXCV* (Boston: Printed by B. Green, for S. Phillips, 1695), [15–16].

31. Capp, *English Almanacs*, 206; Stowell, *Early American Almanacs*, 53–62. Francisco Guerra states that at least 90 percent of colonial American almanacs contained the Anatomical Man. See his "Medical Almanacs of the American Colonial Period," *Journal of the History of Medicine and Allied Sciences* 16 (1961): 237. My own research, albeit based on a much smaller sample, found that Guerra's assessment was accurate only in terms of almanacs printed in the mid-Atlantic and southern colonies. Less than half of the almanacs printed in New England included the Anatomical Man. Astrology was an integral component of Pennsylvania German almanacs. See Louis D. Winkler, "Pennsylvania German Astronomy and Astrology XIII: Health and the Heavens," *Pennsylvania Folklife* 26 (1976): 39–43; and idem, "Pennsylvania German Astronomy and Astrology XIV: Benjamin Franklin's Almanacs," *Pennsylvania Folklife* 26 (1977): 36–43. On popular astrology-related health publications that circulated among early American German communities, see David L. Cowen and Renate Wilson, "The Traffic in Medical Ideas: Popular Medical Texts as German Imports and American Imprints," *Caduceus* 13 (spring 1997): 67–80. See also Christa M. Wilmanns Wells, "A Small Herbal of Little Cost, 1762–1778: A Case Study of a Colonial Herbal as a Social and Cultural Document" (Ph.D. diss., University of Pennsylvania, 1980).

32. *The American Almanack for the Year . . . 1715* (New York: Printed and Sold by William Bradford, [1714]), [20].

33. *Boston Almanack for the Year . . . 1692* (Boston: Printed by Benjamin Harris and John Allen, 1692), 17–18.

34. *Tulley 1689. An Almanack for the Year . . . MDCLXXXIX* (Boston: Printed by Samuel Green, 1689), [16].

35. Travis, *An Almanack of Coelestial Motions and Aspects . . . 1709* (Boston: N. Boone, [1708]), quoted in McCarter, "Of Physick and Astronomy," 157.

36. Ibid., 33–42. In the calendar column designating the moon's place in the zodiac, the part of the body influenced by the moon's place on a particular day was represented in three basic formats: use of the zodiacal symbol representing the part of the body (e.g., the lion denoting the heart), the word denoting the part or organ (e.g., "heart"), or either the zodiacal symbol or word along with a number indicating the place or degree in the sign (1–30 degrees) in which the moon was located.

37. In note 5, I commented on skewed percentages due to differences between general and specialized almanacs. This is also the case with percentages concerning almanacs that included the Anatomical Man. The 50 percent statistic I mention above is skewed by regional differences as well as by the contents of specialized almanacs. While only 20 percent of New England almanacs included the Anatomical Man, 82 percent of mid-Atlantic region almanacs, 76 percent of southern almanacs, and 85 percent of western almanacs included the Anatomical Man. That fewer New England almanacs included the Anatomical Man is due largely to the fact that seventeenth-century Harvard almanacs excluded astrological information.

38. *Poor Richard . . . An Almanack for the Year . . . 1733* (Philadelphia: Printed and Sold by B. Franklin, 1733), [3]. Franklin did not include the Anatomical Man in the 1748 *Poor Richard* and offered no explanation for the omission. Louis Winkler suggests that the woodcut of the figure Franklin had been using had worn out. Winkler's evidence for this claim is that in the 1749 almanac Franklin used a larger, improved cut. Winkler, "Pennsylvania German Astronomy and Astrology XIV," 38–39. Elizabeth Carroll Reilly, in *A Dictionary of Colonial American Printers' Ornaments and Illustrations* (Worcester, Mass.: American Antiquarian Society, 1975), 443–56, presents the various artistic renderings of the Anatomical Man used by colonial almanac makers.

39. *The Virginia Almanack for the Year . . . 1768* (Williamsburg: Printed and Sold by Purdie and Dixon, [1767]), [3].

40. *An Astronomical Diary: Or, Almanack for the Year . . . 1764* (Boston: Printed and Sold by R. and S. Draper, [1763]), [2]. Despite his trust in the moon's power over terrestrial matters, Ames seldom inserted the Anatomical Man in his almanacs because he believed that the image represented superstition. For biographical information concerning Ames, see Briggs, *Essays*. Benjamin Franklin's *Poor Richard* almanac was one of the most successful (and most imitated) of its kind in colonial America, reaching a print run of 10,000 annually. But the circulation figures for Franklin's almanac paled in comparison with those for the almanacs of Nathaniel Ames, father and son, which numbered between fifty and sixty thousand annually. C. William Miller, "Franklin's *Poor Richard Almanacs:* Their Printing and Publication," *Studies in Bibliography* 14 (1961): 98; Stowell, *Early American Almanacs*, x.

41. Clough, *The New-England Almanack for 1703* (Boston: B. Green and J. Allen, [1702]), quoted in Briggs, *Essays*, 61.

42. Taylor, *Pensilvania. 1746. An Almanack, and Ephemeris for 1746* (Philadelphia: William Bradford, [1745]), quoted in John Philip Goldberg, "The Eighteenth-Century Philadelphia Almanac and Its English Counterpart" (Ph.D. diss., University of Maryland, 1962), 145.

43. *Johnson and Warner's Almanac, for the Year 1812* (Philadelphia: Published by Johnson & Warner, [1811]), [19].

44. *The Mechanics' Almanac, and Astronomical Ephemeris, for 1825* (Boston: Stone & Fovell, [1824]), [2].

45. Leeds, *The American Almanack for 1725* (New York: William Bradford, [1724]), quoted in Stowell, *Early American Almanacs*, 22.

46. John Butler asserts that early American almanacs were important sources for occult practices, especially astrology, and thus provide strong evidence that these practices—specifically astrology—survived into the nineteenth century. Butler, "Magic, Astrology, and the Early American Religious Heritage," 328, 334. A history of astrology in early America has yet to be written. Butler's article and his *Awash in a Sea of Faith: Christianizing the American People* (Cambridge: Harvard University Press, 1990), along with Hall's *Worlds of Wonder*, and William D. Stahlman's "Astrology in Colonial America: An Extended Query," *William and Mary Quarterly*, 3d. ser., 13 (1956): 551–63, are useful as far as they go.

47. *Trattato de le vita sobria . . .* (Padua, Italy: Gratioso Perchacino, 1558) was the first edition of Luigi Cornaro's much reprinted and often translated work *Discourses on the Sober and Temperate Life*. Very popular in early America, the first edition of Cornaro's work to be issued in this country was entitled *The Probable Way of Attaining a Long and Healthful Life* (Portsmouth, N.H.: George Jerry Osborne, 1788). See William B. Walker, "Luigi Cornaro, a Renaissance Writer on Personal Hygiene," *Bulletin of the History of Medicine* 28 (1954): 525–34; and Gerald J. Gruman, "The Rise and Fall of Prolongevity Hygiene, 1558–1873," *Bulletin of the History of Medicine* 35 (1961): 221–29. George Cheyne's *An Essay on Health and Long Life* (London: George Strahan and J. Leake, 1724) was first issued in the United States in 1813. See Anita Guerrini, *Obesity and Depression in the Enlightenment: The Life and Times of George Cheyne* (Norman: University of Oklahoma Press, 2000); Henry R. Veits, "George Cheyne, 1673–1743," *Bulletin of the History of Medicine* 23 (1943): 435–52: and Lester King, "George Cheyne, Mirror of Eighteenth-Century Medicine," *Bulletin of the History of Medicine* 48 (1974): 517–39. See also Ginny Smith, "Prescribing the Rules of Health: Self-Help and Advice Manuals in the Late Eighteenth Century," in *Patients and Practitioners: Lay Perceptions of Medicine in Pre-Industrial Society*, ed. Roy Porter (Cambridge: Cambridge University Press, 1985), 249–82. On the "nonnaturals," see Antoinette Emch-Deriaz, "The Non-Naturals Made Easy," in *Popularization of Medicine*, ed. Porter, 134–59; Peter H. Niebyl, "The Non-Naturals," *Bulletin of the History of Medicine* 45 (1971): 486–92; Saul Jarcho, "Galen's Six Non-Naturals: A Bibliographical Note and Translation," *Bulletin of the History of Medicine* 44 (1970): 372–77; and L. J. Rather, "The 'Six Things Non-natural': A Note on the Origin and Fate of a Doctrine and a Phrase," *Clio Medica* 3 (1968): 337–47.

48. The most popular domestic medical book in early America was William Buchan's *Domestic Medicine*. According to Charles Rosenberg, the key factor in Buchan's extraordinary success was that he combined in one book the regimen and longevity tradition with the home treatment tradition. Charles E. Rosenberg, "Medical Text and Social Context: Explaining William Buchan's *Domestic Medicine*," *Bulletin of the History of Medicine*

57 (1983): 22–42. Both Wesley's and Culpeper's works were popular in early America. Wesley's career is discussed in G. S. Rousseau, "John Wesley's *Primitive Physic* (1747)," *Harvard Library Bulletin* 16 (1968): 242–56; and Samuel J. Rogal, "Pills for the Poor: John Wesley's *Primitive Physick*," *Yale Journal of Biology and Medicine* 51 (1978): 81–90. On Culpeper, see Olav Thulesius, *Nicholas Culpeper: English Physician and Astrologer* (New York: St. Martin's Press, 1992). See also David Cowen, "The Boston Editions of Nicholas Culpeper," *Journal of the History of Medicine and Allied Sciences* 11 (1956): 156–65.

49. See note 5 above.

50. A concise examination of traditional therapeutic assumptions is Charles E. Rosenberg's "The Therapeutic Revolution: Medicine, Meaning, and Social Change in Nineteenth-Century America," in *The Therapeutic Revolution: Essays in the Social History of American Medicine*, ed. Morris J. Vogel and Charles E. Rosenberg (Philadelphia: University of Pennsylvania Press, 1979), 3–26. See also John Harley Warner, *The Therapeutic Perspective: Medical Practice, Knowledge, and Identity in America, 1820–1885* (Cambridge: Harvard University Press, 1986); J. Worth Estes, "Therapeutic Practice in Colonial New England," in *Medicine in Colonial Massachusetts, 1620–1820*, ed. Philip Cash, Eric H. Christianson, and J. Worth Estes (Boston: Colonial Society of Massachusetts, 1980), 289–384; George E. Gifford Jr., "Botanic Remedies in Colonial Massachusetts, 1620–1820," in ibid., 263–88; and Kay K. Moss, *Southern Folk Medicine, 1750–1820* (Columbia: University of South Carolina Press, 1999). J. Worth Estes's *Dictionary of Protopharmacology: Therapeutic Practices, 1700–1850* (Canton, Mass.: Science History Publications, 1990) provides a concise introduction to the therapeutic practices of both laypersons and trained physicians during the period covered by his work. See Vivian Nutten's "Medicine in the Greek World, 800–50 BC," in *The Western Medical Tradition, 800 BC to AD 1800*, ed. Lawrence I. Conrad et al. (Cambridge: Cambridge University Press, 1995), 23–25, for an introduction to humoralism. Roy Porter's "Lay Medical Knowledge in the Eighteenth Century: The Evidence of the *Gentleman's Magazine*," *Medical History* 29 (1985): 138–68, is a useful examination of what little difference there was between lay and professional medical knowledge in the eighteenth century.

51. The therapeutic theories of Paracelsus are discussed in Andrew Wear, "Medicine in Early Modern Europe, 1500–1700," in *Western Medical Tradition*, ed. Conrad et al., 310–16; Charles Webster, "Alchemical and Paracelsian Medicine," in *Health, Medicine, and Mortality in the Sixteenth Century*, ed. Webster, 301–34; Allen G. Debus, *The Chemical Philosophy: Paracelsian Science and Medicine in the Sixteenth and Seventeenth Centuries* (New York: Science History Publications, 1977); and Walter Pagel, *Paracelsus: An Introduction to Philosophical Medicine in the Era of the Renaissance* (New York: S. Karger, 1958).

52. *The Wilmington Almanack . . . for . . . 1787* (Wilmington, Del.: Printed and Sold by James Adams, [1768]), [3]. See also J. Worth Estes, "Dropsy," in *The Cambridge World History of Human Disease*, ed. Kenneth F. Kiple (Cambridge: Cambridge University Press, 1993), 689–96. Saltpeter or sal niter (potassium nitrate) was used as a diuretic, resolvent, antiseptic, mild cathartic, and diaphoretic. Estes, *Dictionary of Protopharmacology*, 137.

53. *Farmer's Almanac, for the Year 1824* (New York: Published by James A. Burtis, [1823]), [18]. *Petroselinum* (root and seed of parsley), *Raphanus rusticanus* (root of horse radish), *Sinapi alba* or *nigra* (white and black mustard seed), and *Juniperus communis*

(berries and tops of juniper) contained diuretic properties. Estes, *Dictionary of Protopharmacology*, 108, 150, 163, 178.

54. Dysentery is an inflammation of the large intestine characterized by loose stools containing blood and mucus and by painful and unproductive attempts to defecate (tenesmus). Although diarrhea may have been confused with dysentery during the period covered in this chapter, early American references to "bloody flux," according to K. David Patterson, usually referred to true dysentery. Dysentery may be caused by an ameba, *Entamoeba histolytica*, or by several species of bacteria, especially the genus *Shigella*. Patterson states that generally it is impossible to determine from various early American accounts whether amebic or bacillary dysentery was involved. K. David Patterson, "Dysentery," in *Cambridge World History of Human Disease*, ed. Kiple, 696; idem, "Amebic Dysentery," in ibid., 568–70; idem, "Bacillary Dysentery," in ibid., 604–6.

55. *Poulson's Town and Country Almanac, for the Year . . . 1797* (Philadelphia: Printed and Sold by Zachariah Poulson, Junior, [1796]), [40]. Castor oil (*Recinus communis*), an oil expressed from seeds of castor bean, was used as a cathartic. Whereas James Thacher viewed castor oil as a "gentle and useful" cathartic, Christopher Sauer claimed that it was an "overly violent emetic" that should "never be used internally, except by those who are definitely known to possess extraordinary strong constitutions." Glauber's salt, or sodium sulfate, was named for Johann Rudolph Glauber, a German physician and chemist, who introduced the remedy in the middle of the seventeenth century. The salts were used for their cathartic and diuretic properties. Ladanum, an aromatic gum resin, was used as a stomachic (a medicine that warms and strengthens the stomach). Whey, or lac, a term used for milk, was, according to J. Worth Estes, used to promote all natural secretions. Estes, *Dictionary of Protopharmacology*, 90, 111–12, 166, 205; James Thacher, *The American New Dispensatory* (Boston: Printed and Published by T.B. Wait and Co., 1810), 195–96, 261–62; *Sauer's Herbal Cures: America's First Book of Botanic Healing 1762–1778*, trans. and ed. William Woys Weaver (New York: Routledge, 2001), 87–88.

56. *Smith & Forman's New-York & New-Jersey Almanac, for the Year . . . 1816* (New York: Printed and Sold . . . by Smith & Forman, [1815]), [26]. The rhubarb (*Rheum officinale*) referred to in this remedy is not to be confused with the culinary rhubarb used today. It is a related species that is highly toxic. Rhubarb was used as a cathartic. Saffron (*Crocus sativus*) was used as an aromatic, cordial, and narcotic. Christopher Sauer claimed that saffron was a "very effective powder for the bloody flux." Cardamom (*Eletteria cardamomum*) was used as an aromatic, diuretic, and diaphoretic. Estes, *Dictionary of Protopharmacology*, 39, 56; *Sauer's Herbal Cures*, 85–86, 260–61, 270.

57. *The North-Carolina Almanack for 1816* (Raleigh: A. Lucas, [1815]), quoted in Moss, *Southern Folk Medicine*, 69. Cream of tartar (or cremor tartar), powdered sodium potassium tartrate, was used as a cathartic and a diuretic. Ipecacuanhu was used primarily as a mild emetic and, in some cases, as a diaphoretic. Jalap was considered by James Thacher to be "an efficacious and safe purgative." Jalap's use as a diuretic was espoused by Christopher Sauer, who claimed that it was "an excellent medicine for operating against superfluous watery humors." The combination of jalap and cream of tartar produced a strong purgative. Estes, *Dictionary of Protopharmacology*, 55, 104, 106; Thacher, *American New Dispensatory*, 113, 135–36, 220; *Sauer's Herbal Cures*, 177–78, 181.

58. My remarks on fever rely heavily on Leonard G. Wilson, "Fever," in *Companion Encyclopedia of the History of Medicine*, 2 vols., ed. W. F. Bynum and Roy Porter (Lon-

don: Routledge, 1993), 1, 383–93. See also the chapter entitled "Of Fevers" in Lester S.
King, *The Medical World of the Eighteenth Century* (Huntington, N.Y.: Robert E. Krieger
Publishing, 1971). A useful general treatment of the history of fevers is W. F. Bynum and
V. Nutten, ed., *Theories of Fever From Antiquity to the Enlightenment* (London: Wellcome
Institute for the History of Medicine, 1981). For early nineteenth-century observations
on fever, see Leonard G. Wilson, "Fevers and Science in Early Nineteenth-Century Med-
icine," *Journal of the History of Medicine and Allied Sciences* 32 (1978): 386–407; and Dale
C. Smith, "Quinine and Fever: The Development of the Effective Dosage," ibid., 31
(1977): 343–67.

59. *Thomas's Massachusetts, New-Hampshire and Connecticut Almanack, for . . . 1817*
(Worcester, Mass.: Printed and Sold by Isaiah Thomas, [1780]), [21].

60. Briggs, *Essays*, 319, 385. Nathaniel Ames Sr., founder of *An Astronomical Diary,
or, an Almanack,* compiled the annual from 1726 until his death in the summer of 1764.
His son, Nathaniel Jr., carried on the almanac from 1765 to 1775.

61. *The Columbian Calendar, or New-York and Vermont Almanack, for the Year . . .
1817* (Troy: Printed and Sold by Francis Adancourt, [1816]), [8].

62. *Isaiah Thomas's Massachusetts, Connecticut, Rhode-island, Newhampshire & Vermont Almanack . . . for . . . 1799* (Worcester, Mass.: Printed by . . . Isaiah Thomas, [1798]),
34, 38–39.

63. *Phinney's Calendar or Western Almanac, for the Year . . . 1855* (Bridgeport, Conn.:
Printed for J. Barber & Son's, [1851]), [26]; *Middlebrook's New-England Almanac, for the
Year . . . 1852* (Bridgeport, Conn.: Printed for J. Barber & Son's, [1851]), [26]. See Charles
E. Rosenberg, "Body and Mind in Nineteenth-Century Medicine: Some Clinical Origins
of the Neurosis Construct," *Bulletin of the History of Medicine* 63 (1989): 185–97, on how
ideas about the relationship between body and mind evolved prior to and during the
nineteenth century.

64. For a review of Henry H. Porter's publishing career, see Thomas A. Horrocks,
"Promoting Good Health in the Age of Reform: The Medical Publications of Henry H.
Porter of Philadelphia, 1829–1832," *Canadian Bulletin of Medical History* 12 (1995):
259–87; and idem, "'The Poor Man's Riches, The Rich Man's Bliss': Regimen, Reform,
and the *Journal of Health*, 1829–1833," *Proceedings of the American Philosophical Society*
139 (1995): 115–34.

65. *The Medical and Agricultural Register,* published in Boston and edited by Daniel
Adams, was the first American medical journal published for the layperson. It was issued
in twenty-four parts from December 1806 through December 1807.

66. Bell and Condie criticized not only alternative medical systems, such as Thomsonianism and homeopathy, but also domestic medical practice. They believed that therapeutic advice offered in domestic medical texts—and, by implication, similar advice
that appeared in many almanacs—was "little short of quackery." Horrocks, "Poor Man's
Riches," 132.

67. Ibid., 123.

68. Alex Berman and Michael A. Flannery, *America's Botanico-Medical Movements:
Vox Populi* (New York: Pharmaceutical Products Press, 2001); idem, "The Botanical
Movements and Orthodox Medicine," in *Other Healers: Unorthodox Medicine in America,*
ed. Norman Gevitz (Baltimore: Johns Hopkins University Press, 1988), 29–51; John S.
Haller Jr., *The People's Doctors: Samuel Thomson and the American Botanical Movement,
1790–1860* (Carbondale: Southern Illinois University Press, 2000); Daniel J. Wallace,

"Thomsonians: The People's Doctors," *Clio Medica* 14 (1980): 169–86; Alex Berman, "The Thomsonian Movement and Its Relation to American Pharmacy and Medicine," *Bulletin of the History of Medicine* 25 (1951): 405–28, 519–38. It is worth noting that none of these sources cite any of the various Thomsonian almanacs. See also J. Worth Estes, "Samuel Thomson Rewrites Hippocrates," in *Medicine and Healing,* ed. Peter Benes (Boston: Boston University, 1990), 113–32.

69. *Thomson's Almanac, Calculated on an Improved Plan, for the Year 1842* (Boston: Printed and Published for Dr. Samuel Thomson, [1841]), [17].

70. *Thomson's Almanac for the Year 1843* (Boston: Printed and Published for Dr. Samuel Thomson, [1842]), 3.

71. The health reform movement of the 1830s and 1840s is covered by John B. Blake, "Health Reform," in *The Rise of Adventism: Religion and Society in Mid-Nineteenth-Century America,* ed. Edwin S. Gaustad (New York: Harper & Row, 1974), 30–49; Stephen Nissenbaum, *Sex, Diet, and Debility in Jacksonian America: Sylvester Graham and Health Reform* (Westport, Conn.: Greenwood Press, 1980); James C. Whorton, *Crusaders for Fitness: The History of American Health Reformers* (Princeton, N.J.: Princeton University Press, 1982); and Harvey Green, *Fit for America: Health, Fitness, Sport, and American Society* (New York: Pantheon Books, 1986).

72. See Charles E. Rosenberg, "Catechisms of Health: The Body in the Prebellum Classroom," *Bulletin of the History of Medicine* 69 (1995): 175–97, for a concise and insightful analysis of early-nineteenth-century anatomy and physiology textbooks. See Carl F. Kaestle, *Pillars of the Republic: Common Schools and American Society, 1760–1860* (New York: Hill and Wang, 1983), for a useful overview of the rise of the common school movement in America.

73. *The Health Almanac, for the Year . . . 1842* (New York: Health Depository, [1841]), [18].

74. *The Health Almanac, for the Year . . . 1844* (New York: Sexton & Miles, [1843]), [46].

75. Few texts—learned or popular—containing astrology were printed in America before the middle of the nineteenth century. Moreover, the main focus of these publications was not astrology but fortune telling. For examples, see *The Complete Fortune Teller; or, An Infallible Guide to the Hidden Decrees of Fate: Being a New and Regular System for Foretelling Future Events, by Astrology, Phisiognomy, Palmistry, Moles, Cards, Dreams* (Boston: Printed for the Booksellers, 1797); and Chloe Russel, *The Complete Fortune Teller, and Dream Book, by Which Every Person May Acquaint Themselves With the Most Important Events That Shall Attend Through Life. . . . By Astrology-Phsiognomy, and Palmistry . . .* (Exeter: Published by Abel Brown, 1824). Few of the numerous eighteenth- and nineteenth-century popular medical books, except for those of Nicholas Culpeper and William Salmon, and the various editions of *Erra Pater,* contained astrological health information. Astrology experienced a revival in England during the first decades of the nineteenth century. Beginning in the 1840s, the United States witnessed its own astrology boom of sorts, with the appearance of several books and periodicals devoted to the subject. However, these publications, like the fortunetelling texts of an earlier period, emphasized judicial astrology and contained little health advice. Lewis, *Astrology Encyclopedia;* idem, *The Beginnings of Astrology in America: Astrology and the Re-Emergence of Cosmic Religion* (New York: Garland Publishing, 1990); and L. D. Broughton, *The Elements of Astrology* (New York: L. D. Broughton, 1906).

76. In terms of bleeding, John Harley Warner states that American physicians grad-

ually abandoned "heroic depletive" therapeutic measures during the second quarter of the nineteenth century. Physicians were resorting to bleeding as late as the 1850s, but primarily to control pain, not to provide a cure. Warner, *Therapeutic Perspective*, 93–100. However, one does not encounter a similar shift in the therapeutic advice offered by American almanacs issued during the first half of the nineteenth century.

6

Conflict and Self-Sufficiency

Domestic Medicine in the American South

Although mid-nineteenth-century southerners suffered from many of the same afflictions as other Americans, the South has been seen, for good reasons, as a distinctive setting for disease. What we know best about this setting we know on a large scale. For instance, we know something about the extent and impact of certain widespread diseases like the nearly ever-present malaria and the outbreaks of epidemic yellow fever. And, given the longstanding historical interest in slavery, we know a fair amount about the aggregate health of African Americans in bondage.[1]

We know much less, however, about health and disease close up, in terms of the care-giving practices that ordinary southerners of both races used when they confronted sickness and undertook to fight it in the scattered rural communities so typical of the South in this time. Domestic care giving as a context for medical practice—as a matrix of ideas and techniques drawn from family traditions, community lore, advice and recipe books, and the knowledge of various kinds of healers, including orthodox physicians—was a more complicated, fluid phenomenon than we have appreciated. It was simultaneously traditional and innovative. It deferred to the experience of healers from outside the patient's family as well as the experience of family and friends. Indeed, much of domestic care giving relied upon the willingness of families to take decisive bedside action without outside assistance. At the same time, domestic care was shot through with dread and foreboding, a crucial reality of care giving that has not been given much study. It is difficult to capture such a complex experience in retrospect because so much of it was improvised, invented under pressure, and thus not recorded in systematic texts. Yet because the home was where southerners of both races got sick, cared for the sick, and searched for explanations, it is important to try to understand how the domestic world shaped, and was shaped by, sickness and its remedy.[2]

What follows is an exploration of the domestic context for sickness and health, mostly among white southerners in the mid–nineteenth-century South, with particular attention to what people actually did at the bedside and how they explained it. Published books of domestic health advice have informed our sense of how families combated sickness. These books are valuable sources for reconstructing a domestic context for medicine, with their plain-spoken but weighty ideological language, their detailed descriptions of drugs and ailments, and their directions, assumptions, and exhortations, which imply much about the professional, social, and gender characteristics of the books' authors and readers. And yet, interestingly, in none of several major texts circulating in the South is there a sustained picture of care-giving *activity* at the bedside and the way it illuminated the fluid relationship between sick person and healer as the two struggled against malady. Actual bedside work is veiled, even omitted, in these books. In part, I hope to suggest, this is because the shifting and reactive nature of care giving made it far more deeply troubled than domestic advice book authors acknowledged. Indeed, despite the wide range of styles in these books, from terse, matter-of-fact catalogs of ailments and therapies to expansive pronouncements on the art of good living, authors clearly underplayed the dread of disease, thereby highlighting other factors in the care of the sick that perhaps have been given too much emphasis. Interrogating published advice books with the letters, diaries, and memories of women and men who gave home care and received it, I would like to suggest some revisions of our picture of domestic practice on southern plantations that will help us see not only the common-sense adaptability of this practice but also some of the tensions that reveal it for the edgy, contingent thing it was.[3]

One thing in need of revision, I think, is the image of families responding to sickness with an unruffled self-sufficiency. Looking to sell their books, of course, authors might have been expected to compliment their readers' competence. But most authors went beyond a mere bow in this direction. Through the very orderliness of their practical catalogs of symptoms and remedies, and by way of their confident assurances, authors suggested that appropriate care (incorporating their books, of course) was within easy reach of every household. Samuel Jennings, for instance, typically assured his readers that despite the apparent complexity of disease, "the art of healing is not necessarily beclouded with inexplicable mystery"; he added that intelligent caregivers "of all classes" might not only heal their sick but learn something about medicine while doing so. Ordinary people, author Isaac Wright concurred, could supply the care through which "many lives may be preserved . . . and many diseases may be so arrested in the incipient stages of their career." The picture in domestic books is one of a family alone but nonetheless confronting dread disease with great self-sufficiency, equal to the emotional as well as physical demands of attending the sick.[4]

A second image of the domestic scene that needs some refocusing—and is often in tension with the first—is the sense conveyed by many domestic advice books that the home care giving was influenced by sharply drawn conflict among various kinds of healers. Much has been made of the way authors of domestic advice often harshly criticized rival medical sects and each other, particularly in terms of which medicines were best and whose advice was most dependable. Orthodox physicians often came in for the roughest treatment, criticized for the debility wrought by the body-shaking power of their toxic prescriptions as well as for their hubris and elitism. Popular adviser John Gunn, for instance, who apparently had orthodox training himself, nonetheless typically slammed orthodoxy for its "useless technical terms and phrases" and its "fudge and mummery"; for good measure, he also attacked the "impudent pretensions of science and quackery combined." Orthodox physicians counterattacked; one author typically flayed "the ignorant pretender and the assuming quack" and the "destructive ravages of heartless, unrelenting empiricism," by which he meant any nonorthodox style of healing. Some authors were so contentious, and so given to dire warnings about the wiles of their rivals, that it has seemed as if such conflict dominated the atmosphere of the sickroom itself.[5]

The letters and diaries of patients and their families suggest a more nuanced picture. It is not that sectarian conflict and family self-sufficiency did not exist in domestic care giving; certainly they did. If nothing else, the possibility of both was introduced into the context of home health care by the persistent replaying of these themes in domestic books themselves. But neither such conflict nor self-sufficiency was as fundamental or as absorbing as other features of domestic practice less often articulated by domestic authors. Caregivers themselves talked with greater intensity about other concerns in the less focused but more textured venues of their letters and diaries, which in themselves form a kind of informal text on domestic health. These writings reveal not only procedures and remedies but also the significance of the urgent needs and fears crowding in on the group around the bedside. Drawing on the particular frankness and subjectivity of such writing, I will sketch key features of domestic care giving by beginning with how southerners spoke about sickness as a threatening force in their everyday "well" world. This conversational way of watchfully patrolling the boundaries of their health was linked to how they took hold of the body once it was afflicted, seeking to name disease and counter it. Finally, I will suggest how the routines of domestic plantation practice (with special attention to slavery) were shaped not so much by either sectarian conflict or calm self-sufficiency but by the fluid, frightening swirl of events and predispositions beyond anyone's forethought or control.

A Boundless, Fearful Force

Looming largest on the domestic scene was the vast force of disease itself and the fear it engendered, something that advice books avoided or deflected. Some authors touched upon certain troubling aspects of disease, such as its widely known ability to unpredictably change its identity within the body, and the fact that certain symptoms might mask two or more different diseases. But one can feel most authors reassuring patients, steering them away from fear. This put authors in something of a bind, for at the same time that they raised the fearsome prospect that families' "mere experience" might quickly run dry and necessitate bringing in an outside healer, they also maintained, as one put it, that a doctor's "reading and study . . . should not make us disparage or lightly esteem the benefits of [families'] experience." Thus, while not disregarding completely people's visceral response to sickness, advisers tended to speak of it as one ingredient of a given episode of illness, not as an overarching feature of the social experience of disease. Like other advisers, A. G. Goodlett recognized the importance of countering a patient's fear (he even had a prescription for it: "Tea, a little wine, or spirits and water") but shied away from the fear that gripped those caring for the patient.[6]

In contrast to this cautious approach, the parents and spouses at the bedside, and even a number of physicians and other healers who shared in the making of domestic "experience," informally talked a great deal about what most impressed them about disease: its extensive, protean, at-large power. Disease was as intimate as the home, but it also was broad and ultimately beyond knowing—like the weather or the will of God. And thus everyone's sense of sickness was shaped by how this vast force entered into the intimate confines of family life. As children, watching their families mobilize against disease, southerners learned that the world might turn against them; the heat of summer, sudden changes in temperature, and the very clouds in the sky might portend something malign and personal. Talk about this was constant in letters among family and friends, a kind of continuous low hum of misgiving and monitoring. Sallie Price wrote to her friend: "[H]ow I dread the summer. I almost imagine myself shaking [with chills] now." A Mississippi woman saw autumn's "sickness and distress" waiting in the wings of early summer and wrote to her sister, "We must expect attacks from the invisible enemy, but I pray our share may be light." Personal risk was not the only thing to be dreaded. The social chaos of epidemics was greatly feared as well, one woman recalling her community overwhelmed by cholera, with "the bursting of coffins—the indecency—the haste where there was accorded any burial—not because humanity was gone but because it could not be else." Domestic authors were concerned that such powerful images might drive patients over the edge. But for the families who stood sentinel to loss, such talk was both a sign that one was a vigilant, responsible community member and a personal amulet against the terror and loss.[7]

Southerners were eager to attach names to what afflicted them. They did not wait for the physician but drew on a vocabulary of diseases heard about, read about, or seen before, mixing the diagnoses of doctors, family members, and advice books. So one man was satisfied, if not exactly relieved, when he decided that the growth on his face was "what is called a tumer." Louisa Muller determined that her mother's "severe pain in her side" and rash was "what is called Shingles," and another woman was certain after talking to friends that she had been attacked by "something of the Erysipelas on her Feet." Robina Tillinghast recognized "the cough" when it reentered her household each winter. Domestic advice authors inhabited this same world, often giving more than one name for a malady so readers might better recognize it. And, like A. G. Goodlett, they sometimes adopted vernacular forms of writing as the best way to relate what they knew; in his book, Goodlett framed his thoughts about the South's unique varieties of disease in the form of a familiar letter to a friend.

In both advice books and people's letters, attaching names to sickness helped combat the fear that one was alone in one's trial. Names suggested that sickness had an identity apart from a particular sufferer; it had a known "course" that suggested the possibility of control.[8]

And yet, paradoxically, naming diseases also added to their power as living forces. Although authors used names to tamp down fear, taming afflictions by calmly categorizing them, family members often uttered malady's names in a way that evoked mortal dread. Indeed, the way many people summoned the evocative power of names suggests that domestic authors probably knew that the names they offered had the potential for stirring up fear. Vivid names, however familiar, added to the sense that a rampant killer was moving about the countryside at will. To evoke "the Scarlet Fever prevailing about the country," to leave one's home in order to "keep from the ravages of the scarlet fever," or to say of a city that "Scarlet Fever is there" imparted new life to the disease. The names reverberated throughout domestic advice texts and entered into the home remedy notebooks kept by women. The sheer number of names in advice books is matched by women's multiple sightings of clever, protean disease. The volume compiled by one South Carolina wife and mother is typical, giving antidotes for a shocking array of ailments that suggest the terrible disorder inherent in the wide range of possible afflictions: consumption cough, whooping cough, bowel complaint, spasm in the stomach, worms, indigestion, scarlet fever, pleurisy, dropsy, obstruction, piles, sore nipples, and "what is termed *death mould*."[9]

Thus, the vernacular language for sickness, in its proliferation of names, revealed how agile, cunning disease could not be fully caught by anyone's word or deed. Sickness, multiplying through its particular names, ultimately eluded sharp definition, and thus the sheer number of names, flowing from popular conversation into advice books and back again, added to the widespread sense of malady's transformative power. Southerners' scrutiny of their afflictions identified certain dynamics of sickness that puzzled or divided doctors and authors of advice books. For example, most families seem to have taken as a given what we would call contagion—the understanding that a child's schoolmates might "give" her whooping cough—even though domestic authors usually did not stress it and physicians doubted it. More disturbingly, the desire to identify disease underscored its frustrating and fearsome ability to change from one sort of ailment to another. One woman was certain, for example, that "neglected ulcerated sore throat terminates in consumption" and that the bad head cold suffered by another woman's husband already had transformed into "a turn of intermittent fever." In the Boulware family of upcountry South Carolina, malady can almost be seen ricocheting about the room in the father's diary report on his three sick daughters: "Lulu took a chill & had a hard spasm. Mary Jane sick all day. Lulu clear of fever tonight and very lively. Mattie took a violent headache from the excitement caused by Lulu's spasm." Sometimes, no matter a family's watchfulness and caution, signs were misread and clever disease slipped through; what seemed "ordinary" turned out not to be. So, in the case of a twelve-year-old slave boy, his family and owner had known of his headache and violent shivering, but "supposing it to be an ordinary case of chills . . . no attention was paid to the boy until night, when he was found in a comatose state." How often this happened: the killer unmasked too late.[10]

Disease, Faith, and Morality

Because disease seemed thus to be everywhere and implicated in a person's every habit, it was but a short distance from things natural to things moral. For most southerners, slave or free, a person's choices in life were woven directly into both the moral and material worlds of illness. While some domestic authors—Ralph Schenck and John Hume Simons, for instance—were rather prim about mentioning morality, others seemed to relish linking daily habits and opportunistic disease to moral rectitude, and from there to the vast reaches of religious faith. A. G. Goodlett pressed his conviction that "health is the physical result of nicely balanced appetites and passions, and that there exists no power on earth, that can so attune these into harmony, as *Religion*." "Cast your eye around you," he continued, "and say whence have sprung most of the diseases, both mental and corporal, but from lack of this divine guardian of man, *Religion*." Samuel Jennings's view of morality and illness was less comprehensive, but in cases where the patient's bad habits made the illness worse, or where the sick person was "tormented with unnecessary fears," then the alert caregiver would see that attention needed to "be paid to the state of the patient's mind." "Sometimes there will be occasion for the greatest skill in religious and moral concerns" in the doctor's ministrations, he added.[11]

As with their naming of afflictions, domestic authors and families occupied the same moral terrain, in effect staking it out for each other. But the relatively precise articulation of morality, faith, and health in books only begins to suggest the many flexible and apt ways morality and religious faith impressed southerners with how their every move implicated them in their own suffering. Whether a person drank or ate to excess, was foolish in getting a wet head, or ventured out into the sun too late or too early in the day or in the season—all reflected on the quality and weight of a person's judgment and thus character, shaping health into a moral statement, inevitable and revealing. North Carolinian Mary Henderson spent many weeks retrospectively reviewing the care she and her doctors had given to her young son Archibald, who died despite their efforts. Although she reviewed many particular therapies and diagnoses, searching for their flaws, Henderson also believed that it was her own pride in her child that had moved God to chastise her with her son's death. The remedy for her grief, she told herself, was to beg God: "[G]rant me strength to bear up and learn to kiss the rod which smites me so sorely." Physical and moral language flowed together easily, extending disease's scope and power yet again. Thus a Presbyterian minister, in the course of describing crucial differences between Methodists and Baptists in matters of belief, relied on the physical language of humors to make his point: Methodists are more sanguine, while Baptists "have more Phlegm in their temperament." The moral world shaded easily into the corporeal, and the physical world contained shadows of spiritual lessons. A father in Mississippi, writing to bolster his daughter's religious convictions, advised her to strive for faith the size of a mustard seed; then, reminded by the imagery, he told her, "[W]hen the weather is cold . . . take *mustard seed constantly.*"[12]

Early Diagnosis and Preventive Measures

Southerners spoke frankly about the body, sometimes characterizing it in metaphorical ways also found throughout advice books, seeing the body as a temple to be kept from defilement, or as a fortress under assault by disease. More frequently, they used language that was concretely physical or physiological; keeping watch for changes in the body was a constant theme in their letters. One's body might signify its endangerment by growing too swollen, or, more typically, by being not "fleshy" enough, by becoming "nothing but skin and bones." Feeling "very much reduced and exceedingly feeble" as Susan Yandell did in 1836 was one of the most common ways for southerners to describe a sense of impending illness; the threat of sickness caused their bodies to seem smaller or lighter, vulnerable to a too-free contact with the world, especially the weather. One South Carolinian, like many others throughout the South, be-

lieved that the sudden changes typical of early summer weather "are very trying to my constitution" in a predictable way, sure to leave her "body . . . languid
and lifeless." Mary Ann Shields "can scarcely believe . . . my strength is so much
reduced" by wet, warm weather, which, she feared, was laced with sickness.[13]

People's attention to skin temperature and moisture, and to a troubling
sense of general bodily weakness, gave way to close physical examinations.
Family members looked each other over, taking pulses and examining stools,
freely mixing the advice of neighbors, domestic books, and physicians into a
single set of hopeful but fear-tinged measures. One adviser recommended close
attention to the temperature of the body's extremities, recommending, in fact,
that slave owners go to the quarters at the end of each workday and "feel the
hands and feet of negroes" to be sure that they were not taking a chill. A Virginia woman hoped that her household slave Betty "is spared to me, for she has
been desperately ill, & looks worse than anyone I ever saw to be moving about."
Sophie Tillinghast watched her young son Will closely and wrote her husband:
"[Will] does not look well . . . [He] often seems heavy-eyed . . . I have just discovered that there are little lumps each side of Will's throat & the tonsils look
large." Elizabeth Izard one day found her young daughter feverish, troubled by
her teeth, and examined her: "[U]pon examining her mouth I found her gums
& tongue very much swelled & covered with little ulcers the glands of her neck
very much swelled also." There was little shyness about the body among these
mid–nineteenth-century southerners, and their willing scrutiny of it implies a
deep concern about both the consequences of falling ill and the many physical
acts a skilled caregiver should know how to do in order to take the measure of
health.[14]

This widely shared awareness of the vulnerable body as a complex, mutable
landscape of ever-changing signs helps explain why advice books stopped just
short of describing exactly how the caregiver should take the body in hand. Despite their careful descriptions of symptoms and drugs, authors rarely told
their readers *how* to give a patient a medicine or effect a diagnostic procedure;
sets of detailed steps or stages are almost never found in these books. Instead,
authors used a language of prescription that is somewhat remote from the act:
they simply instructed readers to "apply" or "administer" a drug or poultice;
they noted that a given procedure "should be done" and relied on lofty imper-
atives such as "let the patient take" a given medicine. In none of the well-
circulated advice books is there a description of how to extract a painful tooth,
for example, even though extraction frequently was recommended in this most

common affliction. It might be, of course, that authors believed that most domestic caregivers already had mastered the basic skills. But it also might be
that authors did not go into specifics because beneath the surface symptoms
was a world of detail and contingency that beggared description. Venturing
into details would only raise fears and uncertainties unmanageable in the
pages of advice books. Whatever the mix of reasons, close descriptions of bodies and therapeutic acts are as rare in advice books as they are common in
people's own accounts, a contrast that underscores the striking, fear-tinged ea-
gerness of patients and families to talk about bodies on the edge of affliction or
falling into it.[15]

Along these same lines, southerners took much medicine preventively, be-
lieving that bodies were made to stand up to preemptive measures. Not content
with a tea tonic or a glass of port wine before bed, many people dosed themselves with full-bore drugs such as quinine (especially after 1850) and calomel;
"taking a vomit" was not an unusual event in a prudent family. One advice
book, in line with such practice, typically blurred the boundary between orthodox medicine and other sects by recommending a medicine cabinet with "a
phial of laudanum, one of strong essence of peppermint, with a few doses of
calomel or a box of Wyndam (Lee's) pills." Everyone had a favorite preventive.
Thomas Rutherfoord, like other parents and spouses, advised his family to
concentrate on medicines "for the removal of bile," which was widely thought
to regulate the gut's susceptibility to fever. Harriott Pinckney wrote enthusiastically about plum jelly: a few spoonfuls of it "is excellent for the throat . . . [I]
send three mugs of it." Women, slave and free, dusted children's clothing with
crushed chinchona bark to ward off fever in the air, and youngsters stood
naked over hot coals every evening so that seed ticks would jump out of their
hair and into the fire. Slave mothers, especially, hung packets of camphor and
asafetida around the necks of their children to fend off harmful atmospheres.
And so southerners, their lives surrounded by malady, took the body in hand,
and children of both races grew up considering it prudent and desirable to use
amulets and drugs.[16]

Indeed, the axioms of health pronounced by domestic advice books—"Keep
your body open, your feet dry and warm, and your head cool!!" for instance—
seem far too simple and oddly optimistic when compared to the many layers
of preventive substances and charms laid on by families of both races. Everyone knew that preventive measures often failed, and when they did, southerners did not wait for physicians but poured on additional medicines themselves,

sometimes in spite of doctors' disapproval of substances they regarded as harmful or, at best, as masking the effects of orthodox remedies. Agnes MacRae, bedridden with a bad cold, wrote to her brother that he "assuredly would have pitied me" as her relatives laid on with "plasters, camphorated oil, lemon and sugar, glycerine and whiskey, cream of tartar and paregoric." "I never received a greater variety of doses in my life," she concluded. One former slave recalled that in her antebellum life "[d]ey war no doctahs. Jes use roots and bark for teas of all kinds." Another former slave, Fanny Smith Hodges, similarly recalled being saturated with teas when sick, to the point of bursting.[17]

Good Medicine

To think of all this activity primarily as evidence of a household's care-giving self-sufficiency thus obscures the underlying reality that this "domestic" medicine actually was a composite of drugs and advice pulled from a wide range of sources, published and unpublished, orthodox and alternative. People drew recipes from any source they found creditable, experimenting with them and adapting them to their own needs, and the tide of therapy flowed both ways, shaping the means and materials of doctors as well as households. Moreover, underlying this flood of continually reformatted recommendations was the fear that any one of them might fail at any time, so no text or adviser was beyond doubt and all enjoyed a rough democracy of use. No matter the adviser, then, self-medication was the bedrock of this composite culture of healing. Exchanging advice with a friend in North Carolina, one woman thought it was time to consult the *U.S. Dispensatory*, a pharmacological reference work in use among physicians. "You can see the qualities of the sweet spirits of nitre in pages 298 & 299," she wrote, "and those of seneka rattle snake root in p. 206. Should the fever not abate I suppose it will be necessary to give Arthur another purge. Perhaps the Salts or Castor oil will do better than the Jalap." A Methodist minister suffering from a "rising" or boil under his arm could not bear the doctor lancing it and so determined to do it himself by following directions in a book. He "succeeded in effectually opening the bile which so relieved" him. When David Campbell had a spell of influenza, he turned away all doctors and "took a bench out in the yard and stretched himself first on his back then on his front and made Jimmie pour a pitcher of almost scalding water on his head poured slowly to give time to swett out the cold."[18]

Clearly, physicians were not the only kinds of healers who defined a good medicine by its thumping impact on the body. Physicians were criticized for (and some took pride in) their reliance on aggressive procedures. And yet conscientious family members throughout the South pulled down bottle after bottle of powerful drugs from the medicine shelf, brewed up one decoction after another, and poured them into their charges' bodies. Whatever the importance of the ideological battles among orthodox physicians, eclectics, hydropaths, Thomsonians, and other sects in the world of medical politics, it is clear that the distinctions among them blurred in actual sickrooms. Nothing mattered in the heat of practice as much as the push of contingency, need, and urgent experiment. Mary Henderson's sick son did not like Dr. Sill's Wild Cherry Tonic, so she gave him instead "the domestic preparation W. Cherry and Dogwood in Whiskey." When this did not seem to work, much to her alarm, she resorted to the Sill's tonic on top of the other. After three other combinations of medicines, and some consultation with a physician, the drugs finally seemed to work. As Fannie Moore recalled of her childhood as a slave, the children would try to avoid taking medicines forced on them by caregivers; "but twarnt no use," she explained, "granny jes git you by de collar hol' yo' nose and you jes swallow it or get strangled." A body shaken with the chills and heat of fever had to be strongly seized by a therapy its equal. And so although it was unpleasant—even daunting—most people thought it quite proper, as one South Carolina woman put it, to feel "the good effects" of being made "an invalid" by medicines themselves. Another woman knew she had done the correct thing when she felt her body seized not only by the "prostrating power" of her illness but also by "the thrilling effects of sugar of lead," with which she had dosed herself.[19]

The popular culture of drugs and medical regimen thus was a complicated thing, at once traditional and innovative, objectifying yet intensely subjective, based on past practice, hunch, and rumor. Families had their favorite means shaped by long experience. At the same time, all medical practice was fearfully contingent on unexpected turns of timing seen as fate, on the overarching power of malady, and, finally, on the will of God. Caregivers thus worked their healing measures within a kind of paradox of intention—an active submissiveness, an aroused fatalism—that diminished the significance of conflict between rival medical sects. The urgency of care giving and uneasiness about the outcome made disease a puzzle to be solved; and yet, the sense that huge forces, biological and spiritual, were greater than any one "case" pointed all caregivers and the sick to something beyond all worldly puzzles. Southerners' struggle to occupy these two spheres of care giving did not leave much room for bedside argument over the provenance of a particular therapy or partisan schools of medical thought. ·

Disease, Race, and Slavery

There were other conflicts, however, that meant something to everyone around the bedside. Conflicts between the well and the sick, and among domestic caregivers themselves, deserve more attention as revealing where the southern social fabric wore thin. Perhaps most importantly, slave and free on the southern plantation met each other in the heat of bedside care giving, further complicating the rush of advice and remedies, and thus further qualifying what conflict and self-sufficiency might be said to mean in this time and place. Seen from a demographic viewpoint, the plantation as a site for sickness included the worst of slavery's abuses and struggles; staggering numbers of African Americans—especially children—died of sickness brought on or made worse by bondage. From a more personal viewpoint, one found in the memories of both former slaves and masters, the plantation has been portrayed as the scene of cooperation between the races: the mistress's care of faithful slaves, the fondly remembered slave healer or white physician. Though doubtless there were such persons and times, immediate accounts of sickbed efforts suggest a far more conflicted set of relationships. Although cooperation certainly occurred, slave and master also clearly experienced care-giving relationships as—inevitably—another contest of power under slavery. Success and failure in the face of sickness both had a clear racial dimension; remedies and practices fell into categories as something "negro-like" or as something typical of "buckra."[20]

Indeed, it is clear from the writings of both patients and caregivers that conflict at the plantation bedside was far more likely to be between master and slave than between different kinds of healers brought in from the outside. And yet, strikingly, this conflict is not even hinted at in domestic advice books. In fact, slavery and race are next to invisible, even in books explicitly addressed to a southern audience. John Hume Simons is the exception, concluding his book with three pages of "general directions for raising negroes," in which he noted basic guidelines for hygiene, good diet, and adequate housing that every conscientious slave owner should know. In other authors' writings, the fact that black and white mingled in sickness as in health throughout the South is barely alluded to, with a phrase here and there. Speaking generally of the importance of cleanliness and well-ventilated rooms, for example, A. G. Goodlett remarked in passing on the "misfortune that the poor as well as the slaves are constrained from necessity to sleep in low dwellings, and many of them in the same room." With similar, almost offhand brevity, Samuel Jennings, discussing the troublesome disease cholera morbus (a diagnosis portending serious risk to children) noted, "[I]n treating sick colored children, as well as those that are white, if called early and their condition will certainly warrant a safe reaction, we let blood to prevent congestion."[21]

No simple explanation suggests itself for the way race is veiled in advice books that, in other ways, highlighted their southern setting; there is scarcely a hint that southern society was based on race. One explanation of authors' silence is that, intentionally or not, they shied away from portraying bedside scenes under slavery for fear of seeming overly didactic or intrusive with regard to a slave owner's domain, just as they reined in their occasional criticism of women as caregivers. Authors steered away from such inroads not only because they would offend readers but also because they led into the heart of fundamental social relationships far too volatile to be captured in words. Discussing them would raise racial and gender tensions that authors feared unleashing because they had no idea how to resolve them. In a more immediate sense, the silence on race suggests that despite the mid–nineteenth-century debate among theologians and scientists about the origin of the races, and despite the views of a few physicians on supposed biological differences between whites and Africans, such relatively abstract concerns hardly affected bedside relations. Perhaps, therefore, given the problem-solving mission of advice books, the racial mix of bedside relationships was too familiar to be described or even fully seen, and, in any case, the roles of patient and caregiver simply did not turn on distinctions seen as racial. Thus, the qualities Simons ascribed to a plantation slave "nurse" (a "capable . . . faithful and trusty woman") were no different than the qualities desired in any primary caregiver, such as the *"honesty, sobriety* and *fidelity"* author A. G. Goodlett praised generally in nurses.[22]

In contrast, conflict between slaves and masters is present in the letters and other informal writings of southerners, clearly shaping bedside work. Conflict stemmed, first, from certain fundamental differences between the races in the interpretation of disease. There were deep differences between blacks and whites, rarely articulated at length but obviously important, in what constituted the essence of sickness and the powers underlying its control. This is most clearly seen in the races' very different views of mental or spiritual illness with regard to an African sense of "rootwork." For slaves, the power to define disease and proper care was part of the struggle to grasp any power they could under slavery, usually by indirection or in secrecy. Being able to control what was done for a woman in childbirth or a child with fever was another claim on power against the master's superior force. For masters, the sheer otherness of many African American ideas about illness fed an enormous, if repressed, white uneasiness about covert slave knowledge and intentions across a wide range of social encounters.[23]

Especially where spiritual illness was the issue, many slaves believed self-care was mandatory and white doctors useless at best. Indeed, white healers of all sorts had little to say about what we think of as psychosomatic illness. Whites often were amazed or frustrated by African Americans' belief in the power of conjuring, which permitted the conjurer to injure (or protect) across space and time. One slave owner, a physician himself, wrote in exasperation about a slave's pursuit of a cure: Isaac "is gone today to see a negro doctor, 16 miles in the country, to get medicine 'for poison that a negro gave him in Gibson four years ago'!" Although clearly scornful of and amused at such a diagnosis, the doctor "thought it best to humor his superstition," given that Isaac was determined to pursue the only cure he thought would work. At other times, the alien qualities of slave healers snapped into focus as a resource, a new hope. Victoria Thompson recalled that her father, a slave, was recognized as a valuable healer by his master, who called him "Doc" and deferred to his prescriptions despite orthodox physicians' doubts. Doc dispensed certain remedies to the white family but reserved certain others for the exclusive use of black people. For the whites, there were various teas. For his fellow slaves, "He made us wear charms. Made out of shiny buttons and Indian rock beads. They cured lots of things and the misery too."[24]

In addition to the conflict over defining disease, tensions arose, too, because slave owners folded slave illness into their own drive for mastery, making judgments about sick slaves through a haze of preconceptions about "Negro" behavior. Masters frequently assumed, for instance, that a sick slave was feigning illness (as, of course, sometimes was the case), thus tainting their care giving with suspicion. Gertrude Thomas was doing what many masters did when she treated her slave Daniel's fainting spell with cool skepticism, thinking it contrived; he complained of dizziness and she simply watched him fall to the floor, "rather admiring his skill in effecting it so cleverly." Later, though, she admit-

ted: "I was wrong. He really was sick as was proven." Or, like South Carolina
slave owner Thomas Holloway in 1859, masters used unpleasant medicines as
a deterrent to slaves' "bad" behavior, or as punishment. Faced with a woman
named Pricey who "flinched" from working knee-deep in "sloppy" bottom land,
Holloway asked Dr. Joel Berly for a medicine that would "make her sick
enough" so that she would prefer fieldwork, or, as Holloway put it, be "in good
flight for the clods in the morning." And then there was masters' monetary cal-
culation of the health and well-being of human property, unique to the social
relations of slavery and doubtless a source of underlying resentment and anger
between slaves and slave owners. One Virginia slave owner told his physician,
"I have sent Old Bob to come & see if you can do him any good—if you can
without too much expense." Even if Old Bob and other slaves in like circumstances were unaware of their masters' exact instructions, it would not be difficult to figure out by other means that there was a limit to what would be spent
on their care. Thus "care" was defined not by the patient's need, and not even
by the ability to pay, but by masters' assessment of the economic benefit to
them of treating slaves as opposed to, say, doing nothing or selling the afflicted
people.[25]

Thus conflicts over defining disease and seizing the means of relief—of hope
itself—ran along the well-worn channels of mastery and bondage, opening up
deep, personal contradictions of slavery that profoundly shaped illness and
care in the South, just as it shaped other basic features of southern life. Slave
owner Lucila McCorkle, furious at her slave nurse's disregard of her orders, uttered at the bedside of her sick son a slave owner's prayer for both forbearance
and power: for God "to give me grace, to control myself, and then the authority to control others." And yet mixed together with these conflicts was the undeniable fact that slave and master were mutually exposed to the same debility and death, and many times jointly partook of the potent blend of intimacy,
drudgery, and fear at the sickbed. In many whites' accounts, especially, caregivers of both races lay hands upon each other in ways blunt, familiar, or
painful. Slaves and masters bathed one another, spoon-fed children of both
races, forced down the medicines. There are many scenes as brief and as sharp
as Maria Davies's diary entry in which she took note of her severe sore throat,
recounting how both her sister and her slave Bessie "have been treating my
throat with caustic," holding her in a chair, her sister bracing Maria's tipped-
back head and her slave swabbing her seared throat.[26]

Beyond Prevention and Care

Domestic care inscribed in this way, as small scenes of acting against sickness, suggest that beneath advice books' names, remedies, and explanations for illness was a sense of a dreadful sickness that could not be held still. We return once again to southerners' experience of disease as this kind of force, elusive and fearfully contingent on so many things that remained invisible to the people who tried to help. The sudden, unlooked-for appearance of malady in the midst of the most ordinary day was its central fact and mystery, challenging every means of combating it and every story of its passage. This fear, captured by southerners in their informal writing in a way glossed over by advice books, was where all concern and precaution led, and where our historical inquiry might seek to go.

Consider, both as a coda and as a kind of preface to new inquiry, the account written by North Carolinian Mary Henderson in 1854 about her discovery one day that her small son Edward was ill. It was morning, she remembered, and she was in the dining room with her sister and her slave Polly:

> Little Edward came to the dining room door whilst we were at the breakfast table, told Polly to get his breakfast and amused us all with his brightness and playfulness . . . [A]bout 11 o'clock Polly walked through the passage and he bothered her fretting very much she said to him I know you are going to be sick babe for it aint natural for you to be so fretful. Hearing this I arose immediately [and] went to him and as I always [do] when I suspect fever placed my hand upon his forehead and found it burning hot . . . I told sister he had fever and really was so shocked I became appalled, lost all presence of mind . . . I became completely unnerved and beside myself—we had water put on, a tub brought in, mustard and everything preparatory but Sister begged me to send for a doctor.[27]

Finally, she did call a physician, giving her son some tea and bathing him in the meantime. The doctor came, dosing the boy with his own mixtures, but to no avail; Edward died that same day. Henderson, looking back on that morning, wondered at the terrible completeness of it all. All measures failed; Polly's sympathetic words became a dire forecast, and calling upon the doctor became the summons of the boy's death. Henderson sorted through the events as she might search a story for its moral. Perhaps she was to blame; not she, but her slave, had noticed the first symptoms. As the mother and mistress, she was the

one who should have known the signs and then taken down the books and organized the care, but she had been swept away by something too quick and
strong. Henderson, "beside myself," stared her guilt in the face. And yet she
must have known, too, that sickness's power to mutate and disrupt lay far beyond anyone's knowledge and improvisation. There would be a next time, and
she had no choice but to "do as I always [do]" and wait for the wheel to make
its turn.

NOTES

1. See, for example, Todd L. Savitt, *Medicine and Slavery: The Diseases and Health
Care of Blacks in Antebellum Virginia* (Urbana: University of Illinois Press, 1978); Kenneth Kiple and Virginia King, *Another Dimension to the Black Diaspora: Diet, Disease, and
Racism* (Cambridge: Cambridge University Press, 1981); Reginald Horsman, *Josiah Nott
of Mobile: Southerner, Physician, and Racial Theorist* (Baton Rouge: Louisiana State University Press, 1987); Sharla M. Fett, *Working Cures: Healing, Health, and Power on Southern Slave Plantations* (Chapel Hill: University of North Carolina Press, 2002). Two useful
"sampler" collections on southern regionalism are Todd L. Savitt and James Harvey
Young, eds., *Disease and Distinctiveness in the American South* (Knoxville: University of
Tennessee Press, 1988); and Ronald L. Numbers and Todd L. Savitt, eds., *Science and
Medicine in the Old South* (Baton Rouge: Louisiana State University Press, 1989).
2. Childbirth and its relation to women's illness is one aspect of domestic care giving
in the South that has received attention from historians. See Sally G. McMillen, *Motherhood in the Old South: Pregnancy, Childbirth, and Infant Rearing* (Baton Rouge: Louisiana
State University Press, 1990); Sharla Fett, "Body and Soul: African American Healing in
Southern Antebellum Plantation Communities, 1800–1860" (Ph.D. diss., Rutgers University, 1995); Gertrude Jacinta Fraser, *African American Midwifery in the South: Dialogues of Birth, Race, and Memory* (Cambridge: Harvard University Press, 1998).
3. Domestic medical advice books aimed toward a southern (and sometimes a "western") audience include John Hume Simons, *Planter's Guide, and Family Book of Medicine: for the Instruction and Use of Planters, Families, Country People . . .* (Charleston,
S.C.: M'Carter & Allen, 1848); A. G. Goodlett, *The Family Physician, or Everyman's Companion . . .* (Nashville, Tenn.: Printed at Smith and Nesbit's Steam Press, 1838); Alfred M.
Folger, *The Family Physician, being a Domestic Work, written in Plain Style . . .* (Spartan-
burg C.H., S.C.: Z. D. Cottrell, 1845); Isaac Wright, *Wright's Family Medicine, or System
of Domestic Practice . . .* (Madisonville, Tenn.: Printed at the Office of Henderson, Johnson & Co., 1833); Samuel K. Jennings, *A Compendium of Medical Science . . .* (Tuscaloosa,
Ala.: M. J. Slade, 1847); Ralph Schenck, *The Family Physician . . .* (Fincastle, Va.: Printed
by O. Callaghan & W. E. M. Word, 1842); John W. Bright, *A Plain System of Medical Practice, Adapted to the Use of Families* (Louisville, Ky.: Morton & Griswold, for the Author
[1847]). One of the most widely read volumes was John C. Gunn, *Gunn's Domestic Medicine, or Poor Man's Friend . . .* (Knoxville, Tenn.: Printed Under the Immediate Supervision of the Author, 1830). An excellent introduction to this literature is Charles E. Rosen-

berg's introduction to the facsimile edition of Gunn's volume, above (Knoxville: University of Tennessee Press, 1986); Elizabeth Barnaby Keeney, "Unless Powerful Sick: Domestic Medicine in the Old South," in *Science and Medicine in the Old South*, ed. Numbers and Savitt, 276–94. See also Guenter B. Risse, Ronald L. Numbers, and Judith Walzer Leavitt, eds., *Medicine without Doctors: Home Health Care in American History* (New York: Science History Publications, 1977), and Lamar Riley Murphy, *Enter the Physician: The Transformation of Domestic Medicine, 1760–1860* (Tuscaloosa: University of Alabama Press, 1991).

4. Jennings, *Compendium of Medical Science*, [3]; Wright, *Wright's Family Medicine*, ix. See also Goodlett, *Family Physician*, xv.

5. Gunn, *Gunn's Domestic Medicine*, 99, 98; Wright, *Wright's Family Medicine*, ix. The emphasis by historians on broadly general conflict among healers, of course, is well justified by the struggles among various sects for ideological dominance in the free-market atmosphere of nineteenth-century medical care. But ideological battle in public does not necessarily imply conflict around bedsides. For suggestive works highlighting the conflict of ideas and professional values among healers, including domestic advisers, see Paul Starr, *The Social Transformation of American Medicine: The Rise of a Sovereign Profession and the Making of a Vast Industry* (New York: Basic Books, 1982), esp. chap. 3; William G. Rothstein, *American Physicians in the 19th Century: From Sects to Science* (Baltimore: Johns Hopkins University Press, 1972), esp. chaps. 7 and 8; James C. Mohr, *Doctors and the Law: Medical Jurisprudence in Nineteenth-Century America* (New York: Oxford University Press, 1993), esp. chap. 8; Norman Gevitz, ed., *Other Healers: Unorthodox Medicine in America* (Baltimore: Johns Hopkins University Press, 1988), esp. the essays by Susan Cayleff on hydropathy (82–98) and Martin Kaufman on homeopathy (99–123). John S. Haller Jr., *Medical Protestants: The Eclectics in American Medicine, 1825–1939* (Carbondale: Southern Illinois University Press, 1994), esp. chap. 2; Murphy, *Enter the Physician*, chap. 6.

6. Wright, *Wright's Family Medicine*, 61; Goodlett, *Family Physician*, 154. For other instances of authors' acknowledgment of fear, and their attempts to contain it by describing it as either a manageable clinical problem or an interesting but abstract philosophical issue, see, for example, Gunn, *Gunn's Domestic Medicine*, 20–21; Wright, *Wright's Family Medicine*, 25; Jennings, *Compendium of Medical Science*, 55–57.

7. Sallie Price to Robert Lovett, 6 May 1858, Lovett Papers, Special Collections, Robert W. Woodruff Library, Emory University; M. L. McMurran to L. P. Conner, 7 September 1853, Conner Papers, Louisiana and Lower Mississippi Valley Collections, Hill Memorial Library, Louisiana State University (hereafter LSU); Maria Davies diary, 23 July 1854, Special Collections, William R. Perkins Library, Duke University (hereafter Duke).

8. N. W. Whetstone to William Whetstone, 28 July 1850, Whetstone Papers, South Caroliniana Library, University of South Carolina (hereafter SCL); Louisa Muller to Gerhard Muller, 23 September 1852, Muller Papers, SCL; R. Fletcher to Elizabeth Yates, [14?] February 1825, Yates Papers, South Carolina Historical Society (hereafter SCHS); R[obina Tillinghast] to "My Dear Brother," 5 January 1880, Tillinghast Papers, Duke. See also [?] to "My Dear Brother," 5 January 1861, Gaston-Strait-Wylie-Baskin Papers, SCL; Alicia Middleton to Eweretta Middleton, n.d. [probably 1840s], Cheves-Middleton Papers, SCHS. For Goodlett's "letter," see *Family Physician*, 676–81.

9. James Franklin Torrance to James G. Torrance, 28 February 1847, Torrance-Banks

Papers, University of North Carolina, Charlotte (hereafter UNCC); "Antonia" to Louisa Quitman, 1 February 1848, Quitman Papers, LSU; Mary Henderson to "Ann," 15 November 1852, Henderson Papers, Southern Historical Collection, University of North Carolina, Chapel Hill (hereafter SHC); Isabella Porcher plantation prescription book, ca. 1834, Waring Historical Library, Medical University of South Carolina.

10. Maria Davies diary, 14 December 1852, Duke; M. L. McCurran to L. P. Conner, 7 September 1853, Conner Papers; Thomas and Mary Jane Boulware diary, 13 October 1867, Special Collections, Dacus Library, Winthrop University (hereafter Winthrop); Charles Chester, "Four Cases of Cerebro-Spinal Meningitis," *New Orleans Medical and Surgical Journal* 4 (November 1847): 315. For other examples of this kind of talk, which appears throughout people's letters and diaries, see Sarah Bruce to Ann Rutherfoord, [12?] August 1857, Rutherfoord Papers, Duke; David and Emily Harris diary, 17 and 24 August 1860, Winthrop; Lucila McCorkle diary, 15 May 1847, SHC; Mary Kelley to "My dearest old friend," 6 November 1885, Campbell Papers, Duke.

11. Goodlett, *Family Physician*, 202; Jennings, *Compendium of Medical Science*, 223.

12. Mary Henderson diary, 28 October 1855, John Steele Henderson Papers, SHC; Murdoch Murphy diary, 15 July 1816, Alabama Department of Archives and History (hereafter ADAH); C. Quitman to Louisa Quitman, 4 October 1855, Quitman Papers, LSU. Other examples are William Martin to Lunsford Yandell, 2 January 1845, Yandell Papers, Filson Club, Louisville, Ky.; "Eweretta" to Eweretta Middleton, 14 December 1832, Cheves-Middleton Papers. For Mary Henderson and her diary, which among other things includes a remarkable, extended account of domestic medicines, see Steven M. Stowe, "Writing Sickness: A Southern Woman's Diary of Cares," in *Haunted Bodies: Gender and Southern Texts*, ed. Anne Goodwyn Jones and Susan V. Donaldson (Charlottesville: University Press of Virginia, 1997), 257–86.

13. "Your affectionate Sister" to J. A. Berly, [?] June 1871, Berly Papers, SCL; Susan Yandell to Sarah Wendel, [?] January 1836, Yandell Papers; Lucila McCorkle diary, [20 June] 1847, SHC; Mary Ann Shields to Andrew MacCrery, 11 July 1842, MacCrery Papers, LSU. See also Mahala Roach diary, 23 February 1853, Roach and Eggleston Papers, SHC.

14. "Houses of Negroes—Habits of Living, &c.," *Southern Cultivator* 8 (May 1850), quoted in James O. Breeden, ed., *Advice among Masters: The Ideal of Slave Management in the Old South* (Westport, Conn.: Greenwood Press, 1980), 168; N. Coles to Ann Rutherfoord, 10 April 1858, Rutherfoord Papers; Sophie Tillinghast to William Tillinghast, 26 October 1880, Tillinghast Papers; Elizabeth Izard to Alice Izard, 15–17 January 1814 (holograph copy), Cheves-Middleton Papers. See also Mary Henderson diary, 17 July 1855, SHC.

15. Although generally vague about physical procedures, all authors occasionally describe one. Goodlett described placing leeches, for example, Gunn described handling of the body in treating gleet, and Jennings included directions for people caring for accident victims, all in relatively close detail. But these exceptions only highlight the general absence of such concrete descriptions. Goodlett, *Family Physician*, 637; Gunn, *Gunn's Domestic Medicine*, 275; Jennings, *Compendium of Medical Science*, 565–66.

16. *The American Gentleman's Medical Pocket-Book, and Health Adviser* . . . (Philadelphia: James Kay, Jun. And Brother [ca. 1833]), 3; Thomas Rutherfoord to John Rutherfoord, 19 June 1810, Rutherfoord Papers; Harriott Pinckney to Eweretta Middleton, n.d., Cheves-Middleton Papers. The seed ticks and bark jackets are mentioned in John Bre-

vard Alexander, *Reminiscences of the Past Sixty Years* (Charlotte, N.C.: Ray Printing Co., 1908), 226, 465; for an orthodox physician prescribing a "bark jacket" for a child, see Charles Hentz medical diary, 29 December 1860, Hentz Papers, SHC. On the general dosing with quinine for fever, even in the North, see J. Marion Sims to Paul F. Eve, 25 April [18]74, Eve Papers, Tennessee State Library.

17. *American Gentleman's*, 22; Agnes MacRae to Donald MacRae, 19 September 1879, MacRae Papers, Duke; "Fannie Moore," in *The American Slave: A Composite Autobiography*, ed. George P. Rawick (Westport, Conn.: Greenwood Press, [1971–1972], suppl. ser. 1, 1977) (hereafter AS), 15.2, 134; "Fanny Hodges," AS, suppl. ser. 1, 8.3, 1026. See also Em Tillinghast to [William Tillinghast], 9 October 1886, Tillinghast Papers, who wrote that when she was struck by "a bad pain" while staying with friends, the woman of the household immediately "flew 'round . . . heated water for a hot bath, had a mustard plaster & some medicine of hers that see the very idea & said" that she should not get up. On Americans' concern generally for the bowels as the prime site for disease, see James C. Whorton, *Inner Hygiene: Constipation and the Pursuit of Health in Modern Society* (New York: Oxford University Press, 2000).

18. Mary Campbell to "Dear Mother," 16 May 1803 (a note, apparently either written earlier or misdated, on the reverse of her 7 December 1803 letter addressed "Dear Aunt"), Campbell Papers; George Browder diary, 30 and 31 December 1854, in Richard L. Trautman, ed., *The Heavens Are Weeping: The Diaries of George Richard Browder, 1852–1886* (Grand Rapids, Mich.: Zondervan Publishing House, 1987), 91; "Margaret" to Mary Kelley, 25 January 1870, Campbell Papers. See also James Franklin Torrance to James G. Torrance, 28 February 1847, Torrance-Banks Papers. During the nineteenth century, some changes occurred in all of this self-medicating. Certain drugs passed in and out of favor, especially among white southerners. Calomel seems to dwindle in domestic use after the 1840s; quinine and bromides seem to be mentioned more often. But such trends are difficult to map. Commercial patent medicines certainly became more plentiful after the Civil War and seem in some white households to have been employed in place of older homemade remedies. And yet it still is not uncommon to find southerners preparing their own decoctions, teas, and tinctures into the 1880s. In short, there was more continuity than change in the mainstream domestic regimen, in terms of both the substances employed and the expectations people had of self-therapy. Self-remedies among African Americans, especially, through preference and necessity, persisted during and after slavery. And even among whites, the basic purges and tonics—castor oil, ipecac, and whiskey or wine—seemed to have been in as much circulation in the 1880s as earlier. See, generally, Morris J. Vogel and Charles E. Rosenberg, eds., *The Therapeutic Revolution: Essays in the Social History of American Medicine* (Philadelphia: University of Pennsylvania Press, 1979), and John Harley Warner, *The Therapeutic Perspective: Medical Practice, Knowledge, and Identity in America, 1820–1885* (Cambridge: Harvard University Press, 1986). On the persistence of African American folk remedies, see Loudell F. Snow, *Walkin' over Medicine* (Boulder, Colo.: Westview Press, 1993). For various kinds of southern ailments and remedies, see Kay K. Moss, *Southern Folk Medicine, 1750–1820* (Columbia: University of South Carolina Press, 1999).

19. Mary Henderson diary, 5 September 1855, SHC; "Fannie Moore," AS, 15.2, 134; "Sister" to Eweretta Middleton, 30 November 1832, Cheves-Middleton Papers; Zilla Brandon diary/memoir, 21 June 1860, 235, ADAH. See also Susan Yandell to David Wendel, 18 June 1835, Yandell Papers. James Reed Branch wrote to his eight-year-old daugh-

ter Sallie that he was sorry to hear that she was sick and "had to take that bad physic Cod liver oil"; "I know it tastes and smells badly," he wrote, "but then you must try and take it like a good girl—not make a fuss about it and when I come home I give you a handsome present." James Reed Branch to Sallie Branch, 1 April 1866, James Reed Branch letter, Virginia Historical Society (hereafter VHS).

20. "Buckra" was a colloquial—and pejorative—term for poor whites. For brief and relatively flat accounts by masters of their slaves' deaths, see, for example, Jane Campbell Harris Woodruff, "Narrative of Jane Campbell Harris Woodruff: an Autobiography," [1829], 80–81, TS, Clarkson Family Papers, UNCC; Jane Barr to "Dear Brother and Sister," 11 July 1854, Berly Papers; William B. Hamilton to William S. Hamilton, 29 January 1851, Hamilton Papers, LSU. On the other hand, see the concern expressed by Charles Colcock Jones and his physician son Joseph for the elder Jones's slave Niger, in Robert Manson Myers, ed., *The Children of Pride: A True Story of Georgia and the Civil War* (New Haven: Yale University Press, 1972), for instance, the letter from C. C. Jones to C. C. Jones Jr., 21 April 1862; and the sorrow expressed by Samuel Leland over the death of his slave Jacob, Leland diary, 24 February 1853, SCL. Susan Smedes, like other former masters, fondly recalled her servant Mammy Harriet as an ideal caregiver. Smedes, *Memorials of a Southern Planter*, ed. Fletcher Green (Jackson: University Press of Mississippi, 1981), chap. 4. A great many former slaves remarked on mistresses giving care, sometimes with praise, other times with the implication that such work was merely self-interest; see, for example, "Andrew Jackson Fill," AS, suppl. ser. 1, 8.3, 841, and "Lucretia Brown," AS, suppl. ser. 1, 6.1, 267.

21. Simons, *Planter's Guide*, 207; Goodlett, *Family Physician*, 630; Jennings, *Compendium of Medical Science*, 493.

22. Simons, *Planter's Guide*, 208; Goodlett, *Family Physician*, 629. The section in Goodlett's work, "The Nurse's Guide," is an unusually focused discussion of care giving that hints at the difficulties of working with wives and mothers who (in the physician's view) failed to follow his directions or else did foolish things that weakened his carefully constructed therapeutic course. But, in all, Goodlett grits his teeth and does not allow his criticisms to overwhelm the image of his ideal sickroom helpmate; see 629–42. On the southern context of the mid–nineteenth-century theological debate over whether God created one race or several, which increasingly became part of the wider debate over slavery, see Lester D. Stephens, *Science, Race, and Religion in the American South: John Bachman and the Charleston Circle of Naturalists, 1815–1895* (Chapel Hill: University of North Carolina Press, 2000) and Horsman, *Josiah Nott*. Although ordinary doctors rarely wrote about race or racial biology, a handful of physicians did contribute to the debate over whether there were significant "differences" in caring for African Americans. Josiah Nott was one, as was Natchez physician Samuel Cartwright. For a typical piece from this prolific but idiosyncratic writer, see "Report on the Diseases and Physical Peculiarities of the Negro Race," *New Orleans Medical and Surgical Journal* 7 (May 1851): 691–715.

23. This kind of conflict is discussed in Fett, "Body and Soul." For rootwork and conjuring, see Wilbur H. Watson, ed., *Black Folk Medicine* (New Brunswick, N.J.: Transaction Books, 1984); Joseph E. Holloway, ed., *Africanisms in American Culture* (Bloomington: Indiana University Press, 1991); Snow, *Walkin' over Medicine*, esp. chaps. 1, 2, and 7.

24. Lunsford Yandell to Susan Yandell, 9 April 1860, Yandell Papers; "Victoria Thompson," AS, suppl. ser. 1, 12, 322.

25. Virginia Burr, ed., *The Secret Eye: The Journal of Ella Gertrude Clanton Thomas, 1848–1889* (Chapel Hill: University of North Carolina Press, 1990), 18 August 1856 entry, 150; Thomas Holloway to Joel Berly, 12 May 1859, Berly Papers; James Neal to John Mettauer, 1 August 1839, Mettauer Papers, VHS. To some extent, slaves were almost always suspected of feigning their sickness, and slave owners judged the shrewdness of a given doctor by his willingness to use harsh medicine as a means of exposing supposed deceit; see, for example, R. J. Gage, "Plantation Hygiene," *Farmer and Planter* 8 (February 1857): 25–31.

26. Lucila McCorkle diary, 10 October 1847, SHC; Maria Davies diary, 8 August 1852, Duke. See also Mahala Roach diary, 6 March 1853, SHC.

27. Mary Henderson diary, 21 December 1854, SHC.

7

WILLIAM H. HELFAND

Advertising Health to the People

The Early Illustrated Posters

By 1830, the population of the United States approached 13 million, with urban areas—defined by the U.S. Census as those having at least eight thousand residents—growing more rapidly than the rest of the nation. There were a dozen cities with a population of at least twenty-five thousand, and the largest, New York City, boasted about two hundred thousand. This steadily growing urban concentration brought with it many social uncertainties but also opened new possibilities, among which was an enhanced ability to transmit information economically to large numbers of people. Authors, publishers, and a growing number of entrepreneurs saw lucrative opportunities in this urban growth.

For many years, the nation's first settlers had traditionally learned about new ideas and products through word of mouth, sermons, and letters and printed materials from the countries they had left. Newspapers, beginning in 1704, and magazines, dating from the mid–eighteenth century, quickly became important sources for new information as well. "This day is published . . . a new, correct and handsome Octavo edition of Domestic Medicine, or the Family Physician: Being an Attempt to render the Medical Art more generally useful, by shewing people what is in their own Power, both with Respect to the Prevention and Cure of Diseases, by Regimen and Simple Medicines" began an advertisement in the November 23, 1774, issue of the *Pennsylvania Gazette*. The book was one of the early American editions of William Buchan's bestseller. Such notices of books and pamphlets in addition to announcements of health resorts, alternative systems of medicine, varieties of proprietary medicines, and therapeutic novelties filled the columns of newspapers and magazines; in many newspapers these were the primary types of advertisements. Broadsides too carried information, and as urban populations grew, these postings both inside shops and on walls and facades of buildings took on greater significance.

Prior to the 1830s few messages in publications or broadsides were illustrated. One factor restricting the use of illustrations was cost, for unless they were stock woodcuts supplied by the printer, adding images to advertisements was expensive. Second, to equalize what they felt to be unfair competition, editors at times refused to permit the use of images.[1] Because of enduring limitations in printing technology, they also generally refused to provide space beyond the borders of a single column. During periods of paper shortages they even limited the length of an advertisement to a column inch or two. But advances in techniques of wood engraving as well as the advent of commercial lithography in the 1830s—about thirty-five years after the new process had been discovered by Alois Senefelder in Munich—soon enabled advertisers to find the means to escape such onerous restrictions, ushering in a period of innovative developments that continues to this day.[2] Illustrated posters, illustrated broadsides, and large lithographed trade cards, each in either black and white or chromolithography, began to be seen along with older text-only broadsides, providing a rich mixture of promotional materials.

Proprietary medicines and health subjects represented one of the largest, if not the largest, group to seize on the new media to reach the general public, for broadsides and posters afforded advantages that newspapers and magazines could not. Throughout the nineteenth century, "advertising by means of posters and signs had been generally frowned upon; it had been regarded as an unsightly and undignified method of sales promotion used only by manufacturers of patent medicines, who took no heed of such nice considerations . . . It was especially useful in reaching people who never read newspapers or magazines, notably the immigrants who were arriving in swarms from Europe."[3] Several hundred illustrated health broadsides and posters have survived from the 1830–70 period, and many more were probably published for the same promotional purpose.

Illustrated Broadsides

Broadsides were always an important medium for the dissemination of health information; there are records of more than 130 published in the United States before 1820, and this number increased steadily in subsequent years.[4] Early broadsides informed the public about visits of itinerant physicians, lectures, smallpox inoculation, yellow fever, bills of mortality, book and journal proposals, rules for dispensary patients, and, of course, proprietary medicines. Broadsides during this period were generally not illustrated, largely because copperplate or woodcut images added considerably to printers' costs, and the incorporation of intaglio and relief figures along with letterpress presented added complications. Images that were included were generally small and lacked detail, and many clients were content to use headlines with varying typefaces and letterpress to highlight information. But some advertisers and other patrons demanded illustrations, for they felt that they would attract greater attention. After 1830, illustrations appeared more frequently, largely owing to the growing use of lithography, a more versatile and less expensive reproductive process. Lithography permitted the easy incorporation of images with text and obviated the technical difficulties of melding intaglio images and letterpress.

Portraits began to be included in broadsides more frequently during the middle of the nineteenth century. For example, a woodcut engraving of Zadoc Porter, in Quaker dress, took up a large portion of an 1845 broadside for his Poor Mans Curative Sugar Pills and Medicated Bitters, and Dr. C. W. Kierstead included his portrait in an advertisement for the King of All Pain, his name for a product that would cure almost everything.[5] Woodcut and lithographed portraits on broadsides also announced lectures by visiting phrenologists and physicians, either showing the lecturer himself or his cured patients; for phrenologists, a view of the centers of different functions in the brain, or the varieties of types of analyzed heads, often appeared.[6] In an 1847 broadside Dr. Alvin Warren Foss of Limington, Maine, referred to himself as "Nature's Own Physician!" and included his wife in a joint portrait along with extensive text attesting to his skills.[7] The broadside announcing a visit in the 1840s of Siamese twins Chang and Eng included a woodcut of both posing in what appears to be the dress of their native country.[8] Broadsides for veterinary products, directed to the large farming population that existed at midcentury, would generally feature the woodcut of a horse. And in a broadside for Dr. Parmenter's Magnetic Oil, various men are magically able to discard their crutches, a representation adapted from popular prints on the Ages of Man and the Ages of Woman.[9]

Many midcentury broadsides included an illustration that reinforced the text. Just below the large letters "For Men Only!" was a wood engraving of the interior of Dr. Hay's College of Anatomy, showing rows of specimens of many of the "500 objects on view." Similarly, Dr. S. Guthrie of the Boston Electro-Pathy Institute included a woodcut of parts of his "most approved Electrical machines of wonderful power and virtue" in an 1859 broadside announcing his arrival in Boston. There were also two lithographs for Paine's Celebrated Green Mountain Balm of Gilead and Cedar Plaster, one showing the gathering of raw materials for their manufacture and the other presenting a view of Ascutney Mountain in Vermont, near where the products were made.[10]

Lithographed Trade Cards

With the advertisement for Dr. George Stuart's Botanical Syrup and Vegetable Pills, "The greatest Family Medicine in the World," the illustrated broadside began to change and to approach what we would regard today as an illustrated poster.[11] It is a large lithograph, 87 by 57½ centimeters, and includes a considerable amount of print in addition to a view of Stuart's wholesale and retail premises that takes up more than half of the space. It was printed by one of the foremost lithographers in Philadelphia, Peter S. Duval, who had emigrated from France in 1831 and set up his own firm three years later. Lithography, a planographic process using a flat stone on which a design has been created by a crayon or similar waxy substance and that lends itself to large-size publications of almost unlimited quantities, had considerable advantages over woodcuts and engravings, not the least of which were lower costs and longer press runs. Philadelphia, Boston, and New York became centers of a thriving craft beginning in the 1830s, with major efforts devoted to publishing portraits, views, and other works of art that attempted to imitate oil paintings. In addition, there were views of commercial enterprises, known confusingly as trade cards: "Made to advertise commercial and industrial enterprises, these cards were eye-catching pictures, often in color and usually of large size, sometimes of poster dimensions. Printed on paper, they were actually not cards at all. Probably more of them were made for Philadelphia than for any other large center."[12]

To the printers, advertisements such as Stuart's trade card accounted for much of the lithographic work done on commission. In the first half of the century, these prints "often took the form of views of stores and factories. The shopfronts are rich in urban details and show methods of display, while the factory views record the industrialization of New England" as well as other growing areas of the United States.[13] Posted in steamers and public conveyances, posted on walls, and given to customers, the trade cards helped promote business.[14] Not surprisingly, they were often ordered just after the business was opened, moved, or remodeled. Dr. Stuart's advertisement was a good example, and there were a striking number of other wholesale druggists in Philadelphia who also commissioned trade cards.[15] Competition probably spurred on the trade card business, as each wholesaler had its own trade card created. The messages and intent were similar: all aimed to attract customers. These large lithographs present spirited views of both the interiors and facades of a group of merchants who dealt as both retail and wholesale druggists and also sold chemicals; the animated activity going on in some of these prints inspired more than one reader to become a patron. The trade card for Shepherd's Medicines, published by the Philadelphia lithographers Wagner & McGuigan, was a variant on these views in that it showed a ship loading up with five products in the Shepherd's line, leaving Baltimore for points north, south, and west; here the animated scene takes place in front of the ship instead of the shop.[16] These commercial trade cards were black and white, but there were color cards as well, such as Lewis N. Rosenthal's view of the four-story building at 418 Arch Street, Philadelphia, in which Dr. Hoofland's Celebrated German Bitters and Balsamic Cordial were made.[17] The tasteful border surrounding the building lists the ailments Dr. Hoofland's products were intended to cure.

Stock Chromolithographs

For lithographers such as Duval, Wagner & McGuigan, and other publishers of these advertisements, trade cards, as well as other commercial advertisements, were of secondary importance, their primary activity being the production and marketing of chromolithographs. Their goal was to reproduce the look and feel of oil paintings by lithography and thus to develop an expanded market whereby members of the nation's growing middle class could inexpensively furnish their homes with "art." As a writer noted in 1884, "Every . . . lithographer who sends out well drawn, well colored, well composed . . . posters

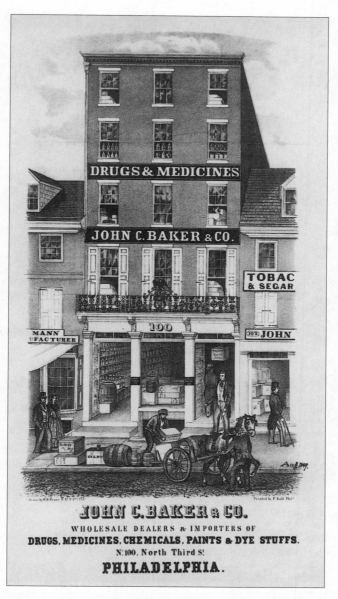

John C. Baker & Co., Wholesale Dealers & Importers of Drugs, Medicines, Chemicals, Paints & Dye Stuffs (Philadelphia: W. H. Rease, artist; F. Kuhl, printer, ca. 1849), lithograph. Courtesy of the Library Company of Philadelphia.

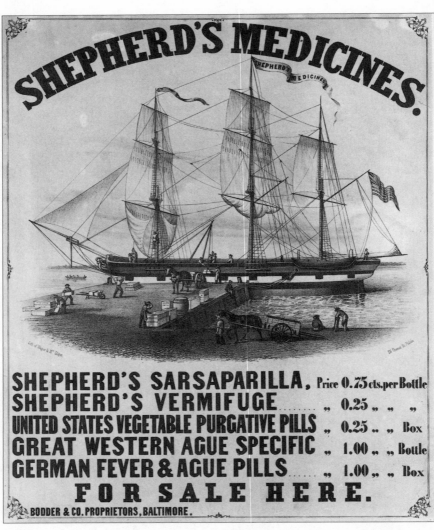

Shepherd's Medicines, Bodder & Co. Proprietors, Baltimore (Philadelphia: Wagner & McGuigan, ca. 1848), lithograph. Courtesy of the author.

to adorn the streets of a large American city is materially assisting in the art education of a Nation."[18]

Once they had published an image, printers frequently attempted to find other places to use it. Often they created advertising posters by combining chromo-lithographs with overprinting. They would either position a lithographic scene within an advertising framework or, the reverse, place the product in a chromo setting.[19] The poster published by the New York firm of Hatch and Company is a good example; in it was recycled the chromolithograph of Molly Pitcher, the 1778 heroine of Monmouth, along with a promotion for Dalley's Magical Pain Extractor. The poster for Harvell's Condition Powders for Horses and Cattle was created in the same way, adding text to a print of the racehorse Dexter *(see Plate 1)*. Posters for Rosenberg's Great East India Horse Invigorator, with its view of horses in a stable; the Old Sachem Bitters, with its Indian chief; and Gainer's Celebrated Spanish Bitters, with its charming scene of several modes of transportation, are similar secondhand uses of stock images.[20] In seeking out stock chromolithographs to offer to marketers, the lithographers tried to find fitting images, and often succeeded. Later in the century, when smaller stock trade cards began to appear, the subjects chosen were often inappropriate, with flowers advertising a dentist's services or small boys promoting cures for headache.

Chromolithographers also published generic borders into which advertising messages could be inserted. The innovative Boston lithographer Louis Prang, a German immigrant who became one of the most successful nineteenth-century lithographers, offered borders containing flags of all nations and arms of all nations. These were for both framing and collecting: "especially adapted for an album collection by children as every Flag or Arm is separately printed and may be cut out by itself." They were also stock items for advertisers.[21] The Boston firm of D. Leland & Co. used Prang's flag series for its Magic Curer; the name and uses of the product were inserted in the center, and the headline "Health to all the World" appeared at the top. Similarly, Whitlock importers used an elaborate chromo border to promote its Wines, Brandies and Liqueurs of Finest Quality for Medicinal Use, as did a rival New York firm, Gilbert & Parsons, to advertise its Hygienic Whiskey for Medicinal Use.[22] The Library of Congress has a copy of a chromolithograph border prepared in 1864 by the Cincinnati lithographers Ehrgott, Ferbriger, and Company to be used for advertising bitters; although the border is elaborately designed with generic bitters bottles, the center is blank, ready to take an advertisement of any product that came along. The sign is about eleven by fifteen inches.[23]

At times, of course, it is difficult to tell whether a poster is a stock item or was especially created for the product itself. "If You Wish Perfect Health, Use the National Bitters," the words on a book held by a young, captivating curly-haired woman, could have been designed specifically for National Bitters or could have been available to all *(see Plate 2)*. Similar questions arise in the case of posters for Dr. Perrin's Fumigator and Laird's Bloom of Youth *(see Plate 3)*.[24]

Unique Posters

The finest midcentury posters are those made specifically for the product or service advertised. Lithographers in Boston called these advertisements show cards: "[U]sed by both retailers and manufacturers, such cards were intended to inform the public of inventory and services by using color pictures in a persuasive and alluring way, often with romantic or humorous imagery. Suitable for window display or countertop ornamentation, show cards were typically seen inside a shop, hung on the walls in a banner style near related merchandise."[25] Though a representative number of these posters have survived, because of extensive use, they are now extremely rare.

Show card imagery reflected a diversity of advertising philosophy. Both J. C. Hurst & Sons and Burnett's Standard Preparations thought it sufficient to display their products, a rather primitive and relatively ineffective method of attracting customers.[26] Factory buildings were another theme, the views being similar in nature to the trade cards of urban shops. William H. Rease, the most active of artists producing these trade cards in Philadelphia, produced two chromolithographs of the Powers & Weightman factories along the Schuylkill River, one from the east and one from the west; each shows the covered bridge in use in the 1860s along with billowing smokestacks *(see Plate 4)*.[27] Powers & Weightman made both medicinal chemicals and industrial chemicals; the firm later became part of Merck and Company.

Not surprisingly, a common approach of early advertisers was to include images of attractive young women. For example, one is simply shown holding a bottle of Laird's Bloom of Youth. In another, a partially nude Indian maiden is botanizing for ingredients that later will be used in the manufacture of Indian Compound of Honey, Boneset and Squills *(see Plate 5)*. In an attractive large poster by the Cincinnati lithographers Gibson & Co., another partially clad woman languidly holds a bottle of Dr. C. W. Roback's Stomach Bitters; she has poured some in a glass, and her relaxed pose may be due to the fact that she has already taken a sip or two *(see Plate 6)*.[28]

For posters made during the early 1860s, military settings also became a theme; Sees' Army Liniment used an image of two Zouaves along with blownup drawings of Sees' products, and Holloway's, a British manufacturer, used a similar scene in promoting its ointment and pills. Schlegel & Co., the New York lithographers, made two additional posters of battle scenes of action during the Civil War, one of a goddess whose presence alone could ensure recovery, and the other showing an officer offering Holloway's panacea to a wounded recruit under the opportunistic headline "Health for the Soldier."[29]

One of the more significant themes in the advertisements was a demonstration of the evils the proprietor's medicine was designed to combat. The Boston artist George Loring Brown used caricature in his early (1837) poster for Phelps' Arcanum, with an angel resting on top of a display of large bottles while joyful patients dance, cripples throw away their crutches, and competitor's products, including Dutch Drops and Swaim's Panacea, are cheerfully destroyed; in the background, trains and ships are ready to distribute the Arcanum to other needy customers. Similarly, in the poster published by Peter S. Duval for Dr. Zane's Antidote, a sure Cure for Drunkenness, the artist imaginatively depicted victims clutching at the large letters in the product's name to save themselves from drowning, but because of the odd juxtaposition of colors chosen, the viewer had to study the poster carefully to understand its significance, and thus it probably did not increase sales.[30] One of the more imaginative treatments of this group of patient's problems was the elaborate poster for Wolcott's Instant Pain Annihilator; its main image is an allegorical illustration of a goddess holding the Annihilator, an act sufficient to ward off the temptations of a frightening demon. While many happy, healthy men and women dance and play under her protection, the borders of the poster include illustrations of the sufferings Wolcott's product was designed to cure, mainly headache and pain.[31]

A related theme presented before-and-after views, long a staple of various types of product advertising and still employed today; several midcentury posters depict changes brought about by the use of the promoted product. Thus an earlier poster for Wolcott's Instant Pain Annihilator is able to do away with pain-producing imps, Dr. Flint's Quaker Bitters appears to be powerful enough to cure cripples, and Dr. C. F. Brown's Young American Liniment is able to significantly improve "All Diseases of Horses and Cattle."[32]

Finally, there are chromolithograph posters that simply show a mother, neighbor, friend, or family member, certainly not a doctor or other health professional, treating the sick. The poster for Dr. McMunn's Kinate of Quinine and Cinchonine has a bedridden woman with a female attendant in a typically American room, with bottles of Dr. McMunn's products at the ready; at the same time three men are gathering a fresh supply of ingredients for future batches from trees conveniently growing outside the door *(see Plate 7)*.[33] Inasmuch as both of the alkaloids in the Kinate are made from the bark of *Cinchona officinalis,* a tree indigenous to Peru and the Andes, it is apparent that the anonymous artist has not given too literal a picture. In addition to Dr. McMunn's Kinate, Kemp's Vegetable Pastilles, Dr. C. V. Girards's Ginger Brandy, Dr. Pierce's Golden Medical Discovery, and Winer's Canadian Vermifuge all made use of this theme of patients being given medicine. In a Winer's Vermifuge poster, a young girl is gently administrating a dose to her doll, a forceful illustration of a domestic scene that designers knew would appeal to women.[34]

Most, but not all, of these compelling posters are chromolithographs, the medium of choice from the 1850s to 1870. Later in the century lithography all but eliminated woodcuts, hand-colored lithographs, wood engraving, and copperplate engraving because of its lower cost. There were some exceptions. The building in Philadelphia in which Dr. Jayne's Family Medicines were manufactured is the subject of an engraving on copper, already an almost extinct technique for commercial printing by the 1850s when it was published.[35] Hand-colored woodcuts and woodblock printed posters also were produced during the years when lithography was in the ascendancy: the before-and-after advertisement for Dr. C. F. Brown's Young American Liniment was relief printed by woodblocks, and another earlier example using this same printing method was the elaborate design made by A. Hanford, a firm of printers in New York City, for the Celebrated Oxygenated Bitters, a Sure Remedy for Dyspepsia, Asthma, and General Debility *(see Plate 8)*. This poster has a portrait of the product's proprietor, George B. Green, surrounded by testimonials from five senators, four congressmen, the president of the Michigan State Bank, and "other prominent gentlemen."[36]

Influence of Posters and Broadsides

Even before the Civil War, there was criticism about the excessiveness, in both product claims and broad distribution, of posters and broadsides, and complaints were most frequently directed at the patent medicine trade, for critics considered such enterprises to be the worst offenders. The author of an 1840 pamphlet complained that nostrum posters were everywhere: on the "walls of our inns—the corners of our streets, and our pumps thereof—the wrecks of burnt, dilapidated buildings, with their standing abutments—the fences enclosing vacant lots in all our cities, if not our small villages, and the decks and cabins of our steamboats be examined, and the columns of our public newspapers."[37]

It was even worse when the war was over. Of course, not all the signs, handbills, and placards were designed to be used outdoors, many of them containing the words "For Sale Here," frequently in large letters. The poster for Dr. Flint's Quaker Bitters went further, with the statement "For Sale Everywhere." The relatively small size of many pre-1870 broadsides, signs, and posters indicates that they were for use in closed surroundings where the viewer could be expected to be nearby.

In contrast to artist-drawn posters, which became popular later in the century, almost all the broadsides and posters published during the middle years of the nineteenth century are unsigned. In the case of chromolithographs, products of commercial firms with large and diversified staffs, this is understandable, for the goal of their publishers was faithful reproduction rather than creative originality. Chromolithography had originated as a handicraft, but by the 1840s it had begun to be a corporate industry. Although probably paid more than the other collaborators, the artist was but one of a number of different craftspeople involved in developing the finished product. By 1890, there were as many as sixteen different individuals involved in chromolithography production.[38] Their products were collective, and no artist's name adorned them. Anonymity may also owe something to the conservatism of craftspeople, who were generally uninterested in graphic experimentation; they did not sign the posters themselves, and it is possible that they did not like the artists to do so either.[39] In many ways the chromolithography firms tended to give greater weight to advertisers' demands than to creative artists'.[40] In the late 1860s, the French artist Jules Chéret, inspired in part by American circus posters, published the first artistic chromolithographs designed to be posted outdoors. His bold, simple, and eye-catching designs could be viewed at a distance on the streets of Paris and seen by thousands, while those inside shops had a more restricted audience. Chéret's influence and that of his fellow artists took some time to develop and did not reach America until the 1890s.

Larger than their American counterparts, European posters also differed in other ways. In European posters, words were integrated with images, while in American posters, originating from the chromolithographers and their stock images, the two were often separate. Such integration made it more difficult for the viewer to quickly grasp the message, a design convention that did not work in favor of the advertiser. And, in the end, the advertiser's objective, whether it was to promote books, educational institutions, or proprietary medicines, was to convince the public to make a purchase. In this, the posters and illustrated broadsides made a distinct contribution, but there were also secondary effects. On one hand, their excesses at times angered the public. On the other hand, the information they carried as well as the images they provided were tools in the growth of a consumer society, a society that became even more sophisticated later in the nineteenth century. According to a New York lithographer of the period, people "do not care so much for black and white as they used to—they want color; as realism seems to prevail, they want in their pictures the color of nature . . . our pictures are many of them for the soap manufacturer, the insurance companies, and the patent-medicine man; but we try in our way to be educators of the people, and to give them good drawing and harmonious coloring. These business patrons of ours who use pictures for advertising purposes know that the public have become fastidious; hence they will only accept good designs."[41]

By the information they provided and the questions they raised, the posters and broadsides of the mid–nineteenth century, with their eclectic mixture of different types of planographic and intaglio media, also helped to enable men and women to involve themselves in their own health care, a persistent social reality that continues to grow in importance in today's intense and omnipresent media environment.

NOTES

1. Frank Presbrey, *The History and Development of Advertising* (New York: Doubleday, Doran and Company, 1929), 232.

2. For a discussion of techniques of printing methods, see Bamber Gascoigne, *How to Identify Prints* (London: Thames and Hudson, 1986).

3. Ralph M. Hower, *The History of an Advertising Agency: N. W. Ayer & Son at Work 1869–1939* (Cambridge: Harvard University Press, 1939), 99.

4. Robert B. Austin, *Early American Medical Imprints* (Washington: National Library of Medicine, 1961).

5. *Mr. Zadoc Porter's Poor Man's Curative Sugar Pills* (New York: 1845), lithograph, Library of Congress; *Dr. C. W. Kierstead's Unrivaled Remedy, the King of All Pain . . .* (Philadelphia: 1863), lithograph, author's collection.

6. *Prof. A. Clapp, Phrenologist, Know Thyself, A Few Days Only, Phrenology and Phys-*

iology (Buffalo, N.Y.: Jewett, Thomas & Co., Electrotypers and Printers, ca. 1850); *Prof. John Logan, Lecture, Humorous and Instructive, Reading Character* (Worcester, Mass.: ca. 1870); *Prof. O. S. Fowler On Phrenology and Physiology* (Worcester, Mass.: ca. 1860); *Prof. Wilson will deliver a Free Illustrated Lecture* (Worcester, Mass.: 1869), all lithographs, American Antiquarian Society; *Know Thyself! Lectures on Phrenology* (New York: ca. 1865), lithograph, author's collection.

7. *Dr. Alvin Warren Foss, Nature's Own Physician* (Saco, Maine: Democrat Press, 1847), lithograph, American Antiquarian Society.

8. *Siamese Twins, For ___ Day only* (New York: J. M. Elliott, Printer, 33 Liberty Street, ca. 1845), lithograph, Ars Medica Collection, Philadelphia Museum of Art.

9. *Read, Reason & Reflect! Dr. Parmenter's Magnetic Oil! Will Cure Rheumatism!* (Albany, N.Y.: Baker Taylor, Printer, 56[?] State Street, ca. 1870), lithograph, New-York Historical Society.

10. *For Men Only! Dr. Hay's College of Anatomy* (Coldwater, Mich.: Kitchel's Liniment Print, ca. 1870), lithograph, author's collection; *Boston Electro-Pathy Institute, Dr. S. Guthrie* (Boston: 1859), lithograph, American Antiquarian Society; *Green Mountain Boys Gathering Materials for Paine's Celebrated Green Mountain Balm of Gilead and Cedar Plaster* (Boston: Forbes & Co., Lith., 159 Washington St., 1868), lithograph, New-York Historical Society; *View of Ascutney Mountain* (Boston: Forbes & Co., Lith., 159 Washington St., 1868), lithograph, New-York Historical Society.

11. *Dr. George Stuart's Botanical Syrup and Vegetable Pills, The greatest Family Medicine in the World* (Philadelphia: P. S. Duval, 1849), lithograph, Library Company of Philadelphia.

12. Nicholas B. Wainwright, *Philadelphia in the Romantic Age of Lithography* (Philadelphia: Historical Society of Pennsylvania, 1958), 2.

13. Sally Pierce and Catherine Slautterback, *Boston Lithography, 1825–1880* (Boston: Athenaeum, 1991), 7.

14. Wainwright, *Philadelphia in the Romantic Age of Lithography*, 5.

15. Including John C. Baker, William Clark, John Horn, Moyer and Hazard, J. & J. Reakirt, George Ridgway, Ritter Cotterell & Ritter, Robert Shoemaker, and Wetherill & Brother. See Wainwright, *Philadelphia in the Romantic Age of Lithography*, numbers 201, 467, 204, 242, 192, 149, 314, 316, 450, all lithographs in the collection of the Library Company of Philadelphia except the lithograph of J. & J. Reakirt (192), which is in the collection of the Historical Society of Pennsylvania.

16. *Shepherd's Medicines*, Bodder & Co. Proprietors, Baltimore (Philadelphia: Wagner & McGuigan, ca. 1848), lithograph, author's collection.

17. *Dr. Hoofland's Celebrated German Bitters and Balsamic Cordial Prepared by Dr. G. M. Jackson* (Philadelphia: L. N. Rosenthal, 1859), chromolithograph, Library Company of Philadelphia.

18. Anonymous, "Street Lithography," *Art Age*, December 1884, 57, quoted in Peter C. Marzio, *The Democratic Art: Pictures for a 19th-Century America* (Boston: David R. Godine, 1979), 131.

19. Marzio, *Democratic Art*, 194.

20. *Dalley's Magical Pain Extractor* (New York: Hatch & Co., 25 William St., 1860), chromolithograph; *Harvell's Condition Powders for Horses and Cattle* (Albany, N.Y.: Charles Van Benthuysen & Sons, 1869), chromolithograph; *Rosenberg's Great East India Horse Invigorator* (New York: Lith. of Sarony, Major, & Knapp, 449 Broadway, 1861),

chromolithograph; *Old Sachem Bitters* (New York: Lith. of Sarony, Major, & Knapp, 449 Broadway, 1859), chromolithograph; *Gainer's Celebrated Spanish Bitters* (Philadelphia: 1868), chromolithograph. All but *Harvell's* (author's collection) are in the Library of Congress.

21. *Prang's Chromo*, January 1868, quoted in Katherine M. McClinton, *The Chromolithographs of Louis Prang* (New York: Clarkson N. Potter, 1973), 54–55.

22. *Leland's Magic Curer* (Boston: Louis Prang, ca. 1870), chromolithograph, American Antiquarian Society; *Benj. M & Edw. A. Whitlock & Co., Wines, Brandies and Liqueurs of Finest Quality for Medicinal Use* (New York: ca. 1865), chromolithograph, American Antiquarian Society; *Gilbert & Parsons, Hygienic Whiskey for Medical Use* (New York: Robertson, Seibert, & Shearman, 1860), chromolithograph, Library of Congress.

23. *Stomach Bitters* (Cincinnati: Eurgott, Ferbriger & Co., 1864), chromolithograph, Library of Congress.

24. *If You Wish Perfect Health, Use the National Bitters*, Schlichter & Zug, Proprietors, Philadelphia (Philadelphia: R. F. Reynolds, 1867), chromolithograph, Library Company of Philadelphia; *Dr. Perrin's Fumigator for Catarrh, for Sore Throat* (Albany, N.Y.: Charles Van Benthuysen & Sons, 1869), chromolithograph, New-York Historical Society; *Laird's Bloom of Youth* (New York: 1860), chromolithograph, Library of Congress.

25. Luna Lambert Levinson, "Images That Sell," in *Aspects of American Printmaking, 1800–1950*, ed. James F. O'Gorman (Syracuse, N.Y.: Syracuse University Press, 1988), 82.

26. *J. C. Hurst & Sons Standard Remedies* (Philadelphia: ca. 1870), chromolithograph, New-York Historical Society; *Burnett's Standard Preparations* (Boston: 1865), chromolithograph, National Museum of American History, Smithsonian Institution.

27. *East View of Schuylkill Falls Laboratory* and *West View of Schuylkill Falls Laboratory*, Powers & Weightman, Manufacturing Chemists, Philadelphia (Philadelphia: William H. Rease, ca. 1860), color lithographs, both in Philadelphia Museum of Art, The William H. Helfand Collection.

28. *Laird's Bloom of Youth* (New York: 1860), chromolithograph, Library of Congress; *Indian Compound of Honey, Boneset and Squills* (New York: Charles H. Hart, ca. 1870), chromolithograph, New-York Historical Society; *Dr. C. W. Roback's Unrivaled Stomach Bitters*, Prince, Walton & Co., Sole Proprietors, Cincinnati (Cincinnati: Gibson & Co., 1866), chromolithograph, Library of Congress.

29. *Sees' Army Liniment* ([New York?]: 1862); *Holloway's Ointment and Pills* posters (New York: Geo. Sculegel Lith., 83 William St., 1862, 1863), both chromolithographs, Library of Congress.

30. *Phelps' Arcanum* (Boston: 1837), lithograph, American Antiquarian Society; *Dr. Zane's Antidote, a sure Cure for Drunkenness* (Philadelphia: P. S. Duval & Son, 1864), Library of Congress.

31. *Wolcott's Instant Pain Annihilator* ([New York?]: 1867), chromolithograph, Library of Congress.

32. *Wolcott's Instant Pain Annihilator, Fig. 1, Demon of Catarrh, . . .* (New York: Endicott & Co., Lith., Beekman St., 1863), chromolithograph, Library of Congress; *Dr. Flint's Quaker Bitters* (New York: ca. 1870), chromolithograph, Ars Medica Collection, Philadelphia Museum of Art; *Dr. C. F. Brown's Young American Liniment* (Philadelphia: Duross Bros., 1861), color woodcut and relief print, Library Company of Philadelphia.

33. *Dr. McMunn's Kinate of Quinine and Cinchonine in Fluid Form and Always Ready for Use* (New York: Thomas and Eno, ca. 1870), chromolithograph, author's collection.

34. *Kemp's Vegetable Pastilles for expelling Worms from the System* (New York: 1857), chromolithograph, Library of Congress; *Dr. C. V. Girards's Ginger Brandy* (New York: 1869), chromolithograph, New-York Historical Society; *Dr. Pierce's Golden Medical Discovery* (Buffalo, N.Y.: 1870), chromolithograph, private collection; *Winer's Canadian Vermifuge, a Speedy and Effectual Remedy for Worms* (New York: 1860), chromolithograph, Library of Congress.

35. *Dr. Jayne's Family Medicines* (Philadelphia: ca. 1850), engraving, Library Company of Philadelphia.

36. *Dr. C. F. Brown's Young American Liniment* (Philadelphia: 1861), Library Company of Philadelphia; *The Celebrated Oxygenated Bitters, a Sure Remedy for Dyspepsia, Asthma, and General Debility,* M. V. B. Fowler, wholesale agent, H. H. Jones, retail agent (New York: A. Hanford, ca. 1846–47), color woodcut and relief print, Philadelphia Museum of Art, The William H. Helfand Collection.

37. William Euen, *A Short Expose on Quackery* (Philadelphia: 1840), 6.

38. Marzio, *Democratic Art,* 149.

39. Victor Margolin, *American Poster Renaissance* (New York: Watson-Guptill Publications, 1975), 17.

40. Harry T. Peters, *America on Stone* (Garden City, N.Y.: Doubleday Doran, 1931), 24.

41. *Art Amateur,* December 1894, 15, quoted in Marzio, *Democratic Art,* 194–95.

8

JEAN SILVER-ISENSTADT

Passions and Perversions

*The Radical Ambition
of Dr. Thomas Low Nichols*

Dr. Thomas Low Nichols hoped to change the world with a book. He did not expect to lose his popular column in the *Water-Cure Journal* over the publication of *Esoteric Anthropology. A Comprehensive and Confidential Treatise On The Structure, Functions, Passional Attractions And Perversions, True And False Physical And Social Conditions, And The Most Intimate Relations Of Men And Women.* This remarkable work, first published in 1853, provoked its readers with a warning: "This is no book for the center-table, the library shelf, or the counter of a bookstore. As its name imports, it is a *private treatise* . . . It is such a book as I wish to put into the hands of every man and every woman—yes, and every child wise enough to profit by its teachings—*and no others.*"[1] Perhaps not surprisingly, the text sold thousands of copies and saw many editions, the latest claiming to constitute a fourteenth printing, published in 1916. In fact, it is still easy to find nineteenth-century copies over the Internet—evidence of the number of copies originally circulated. What made this work so popular? How did it differ from other popular health manuals of its day? What "true conditions" did it sanction? What "perversions" did it address?

The work was written in what Nichols recalled as the mornings of only six weeks.[2] (He was originally a New York City journalist, accustomed to looming deadlines and quick turnaround times.) Dismayed by what he described as the lack of any thorough, accurate, accessible, and comprehensive guide to anatomy, physiology, hygiene, and social science, Nichols sat down at his desk determined to fill this void. As a leader of the health-reform movement and an advocate of water cure, Nichols claimed to have been motivated by duty. "I write, not to get consultations, but to prevent their necessity;" he assured, "not to attract patients, but to keep them away" (6). He already had a working draft: extensive lecture notes that constituted a portion of the core curriculum of the American Hydropathic Institute, a private, unaccredited, and short-lived medical school that Nichols had cofounded with his wife. Because he had been living with the book's central themes for so long, the final draft seemed more received than wrought. At 482 pages, the book was an inch and a half thick and discreet, bound in flexible brown or black muslin in a pocket-sized edition of approximately four by six inches. With nearly seventy anatomical illustrations and the promise of candid guidance, it must have seemed a bargain at one dollar per copy, postage included.

Nichols divided his work into thirty-one chapters. The table of contents suggests nothing particularly unusual. Subjects treated include such general topics as "The Divisions of the Human Body," "Evolution of the Foetus," "Principles of Physiology," "Diseases and Treatment," and "Gestation and Parturition." More specific chapters address diseases of the brain, the respiratory system, the digestive system, and the "generative" system. Space is devoted to "Processes of Water Cure" as well as "Lactation and the Management of Infants." A closing chapter, "On Death," includes excerpts from the reflective letters exchanged between Thomas Jefferson and John Adams on the subject of mortality. With such practical themes, Nichols targeted his writing to those seeking usable knowledge. The popular health movement of antebellum America reflected a general upsurge of interest in the workings of the human body; laypeople sought information that would help them rely less on physicians and have a greater sense of personal control over morbidity and mortality.

As a medical textbook intended for general readers, *Esoteric Anthropology* faced stiff competition. Many mid–nineteenth-century "experts" vied for the role of trusted authority with regard to health maintenance. Although he had earned his formal medical degree from the University of the City of New York, Nichols advocated the water cure (also known as hydrotherapy), an alternative approach to healing that challenged the highly interventionist and "heroic" measures of mainstream medical practice. Regular physicians—"allopathic," their sectarian antagonists termed them—did not control the medical marketplace. Homeopathy, botanic medicine, and water cure represented three of the strongest threats to allopathy; popular health manuals recruited people to their respective philosophies. What made the water cure distinctive was its sole reliance on pure, cold water as a therapeutic measure, coupled with its heavy emphasis on what we would today call "lifestyle choices": what to eat, how much to exercise, what to wear, how to balance work and recreation, and how often to have sex.[3]

Given that the early 1850s saw the publication of numerous books and pamphlets on sex, marriage, hydrotherapy, and self-management generally, it helped that Nichols wrote with style. He characterized the existing market as dominated by "ponderous works on anatomy, dry details of organism, buried in Greek and Latin technicalities, with no more life than the wired skeletons and dried preparations which they describe" (5). *Esoteric Anthropology* could not be described as lifeless. In places, it verged on pornographic. In others, it encouraged open rejection of what most Americans considered the fundamentals of a moral, Christian life. Much of this content appeared in a chapter entitled "Miscellaneous," wedged quietly in the middle of the book, between chapters entitled "Pregnancy" and "Symptoms of Health."

Thomas Nichols knew better than to open his work with explicit provocation. The first hundred pages sensibly offered a grounding in basic anatomy, introduced by this argument: "Man, in his body and his soul, is preeminently a revelation of God; and, in health, is himself the highest known expression of Divine wisdom and love" (14). This description was meant literally. Like many health reformers of his era, Nichols genuinely believed that "When man is in harmony with nature, he is in harmony with God" (12). The conflation of "natural law" with "God's law" created a sturdy bridge between theology and science. It reassured religious readers that physiological study was indeed a blameless pursuit; at the same time, it invited medical empiricists to contemplate the moral implications of health and disease. Ultimately, all readers of *Esoteric Anthropology* were encouraged to reconsider social arrangements that Nichols viewed as "inharmonious." For above all else, Nichols was a social reformer. "This reform of health must be the pivot of all other reforms," he had once written. "So long as people are diseased, nothing can be done for them. A disordered mind in a diseased body makes a bad subject for social reorganization."[4] Hence the appeal of *Esoteric Anthropology* derived not only from the work's medical purpose but from its broader ambitions.

At its core, the book examined sex and power. *Esoteric Anthropology* was indeed a very useful and practical medical textbook, but what made it exceptional—and scandalous—were its radical feminism, its advocacy of free love, and its unabashed descriptions of sex. The work's direct acknowledgment of women's strength and sexuality must have been extremely engaging for readers, whether they responded with sympathy, shock, or contempt. Within the first four months of its release, demand for the "comprehensive treatise" had led to six printings; when sales topped ten thousand copies in only half a year, Nichols contracted with the New York publishing house of Stringer & Townsend for subsequent distribution. Within two years, more than twenty-six thousand copies had been sold—a respectable record for any general health manual of that time.[5]

Nichols intended that his words be read privately, as one might savor an intimate letter. Indeed, he declared the work "written with all the frankness of a private letter, under the seal of professional confidence"—as if in reply to a medical consultation.[6] In promotional blurbs for the work, readers described loaning it to friends and spouses. Because the book was so broad ranging, it left room for readers to recommend the text generally without necessarily sanctioning its more radical views. Even Nichols's wife, prominent lecturer Mary S. Gove Nichols, hedged her endorsement of *Esoteric Anthropology*: "Though it contains some facts which I should not record, and some thoughts which I should never think . . . I recommend it to all, and especially to woman, as a work full of saving knowledge."[7]

In antebellum America, women were being widely encouraged to view themselves as their families' guides to morality, spirituality, and hygiene. Women in the rising middle class absorbed the ubiquitous message that domestic responsibilities included hygienic homemaking, pious childrearing, and dutiful care of a working husband. The management of meals, clothing, sickness, worship, and leisure activities was part of woman's sphere of rightful influence, as she was deemed more pure and less coarse than man. Popular images of domestic motherhood suggested an orderly, peaceful, and refined existence—one devoted to polite arts and parlor entertaining. Like other utopian radicals of his era, Nichols viewed such idyllic scenes with skepticism. While he agreed with the general sentiment "No man can know what a world of delicate tenderness finds its seat in a woman's swelling bosom," he also felt that "this wide difference does not prove that woman was intended to be the slave, the tool, and the victim of man, as she is and has been" (204). Nichols scorned scenes of "blissful" marriage that depended upon women's sexual, emotional, or intellectual sacrifice. "As health is the condition of a true life, the result and sign of health is happiness," wrote Nichols (230). "Hence all unhappiness of every kind, all pain, grief, regret, jealousy, discontent, anxiety, is the result of disease, bodily or mental, in ourselves or others" (230). These words had enormous implications: if misery indicated *disease*, and if women were the rightful managers of household health, then their own discontents necessitated tangible change. It would be medically irresponsible to respond passively. Happiness was what the doctor ordered.

If the pursuit of happiness was futile without the simultaneous pursuit of health, then the personal became political. While few women in 1853 considered the fight for women's rights to be a ladylike pursuit, most everyone believed it highly appropriate for women to educate themselves as to the means of attaining a healthful household. *Esoteric Anthropology* sought to influence these masses. Because it clothed revolutionary perspectives in the gown of preventive health, the book did not require its readers to adopt militant attitudes, but rather maternal ones—to fight for "health," not "equity." Nichols taught that a woman's experiences in the bedroom, for example, could either strengthen or debilitate; her control within that realm, therefore, warranted medical consideration. Private symptoms could indicate the need for public reform. The book made these connections explicit: "It is the part of woman to accept or repulse; to grant or refuse. It is her right to reign a passional queen; to say, 'thus far shalt thou come, and no farther' . . . [T]here is no despotism upon this earth so infernal as that which compels a woman to submit to the embraces of a man she does not love; or to receive even these, when her nature does not require them, and when she can not partake in the sexual embrace without injury" (151).

Nichols referred to indissoluble marriage, to the dependent status of women in the eyes of the law, and to the cultural expectation that wives submit sexually to the wishes of their husbands. He also referred to marital rape—a concept familiar today, but as yet unnamed in 1853, when a husband's right to beat his wife and children went largely unchallenged. "If a woman has any right in this world," insisted Nichols, "it is the right to herself; and if there is any thing in this world she has a right to decide, it is who shall be the father of her children. She has an equal right to decide whether she will have children, and to choose the time for having them" (151). A chapter that had opened with discussion of cellular reproduction in plants quickly wound its way into political rhetoric that even leaders of the women's rights movement found worrisome. For Nichols was advocating "free love," the right of any person to follow his or her "passional attractions."

Although the free-love movement was generally caricatured as a defense of promiscuity and sexual indulgence, such a depiction greatly oversimplified the stakes involved. In fact, *Esoteric Anthropology* explicitly warned against sexual excess: "[N]o one, male or female, ought to average more than once a week" (200). What Nichols understood as the core of free love was a rejection of legal constraints on sexuality. He redefined "adultery" as sex without love rather than sex outside of marriage. Thus a sinful encounter might happen within the bonds of a detested marriage, while a mutually passionate love between two unmarried adults might be expressed without fear of moral reproach. "The true marriage is not a trap in which people are caught," wrote Nichols. "It is a condition of mutual attraction in absolute freedom" (235).

In 1854, the year after he published *Esoteric Anthropology*, Nichols and his wife wrote a lengthy and scathing attack on the institution of traditional marriage, entitled *Marriage: Its History, Character, and Results; Its Sanctities, And Its Profanities; Its Science And Its Facts. Demonstrating Its Influence, As A Civilized Institution, On The Happiness Of The Individual, And The Progress Of The Race.*[8] It was a heartfelt and painstaking elaboration of the political themes embedded in *Esoteric Anthropology*. At a distance, it is easy to see how such strident defense of passional freedom could be confused with licentiousness. Many feared that a liberalization of divorce laws would hurt more women than it would help, leaving wives and children abandoned without adequate financial support. The Nicholses rejected as "fearful" the social structures and strictures that kept women financially dependent on men, like "parasites," and that obliged men to support other wholly capable adults.[9] Comparing marriage to slavery, *Esoteric Anthropology* taught that the healthy expression of God-given passions would fulfill natural law and prevent illness, while existing social conditions reduced women to property and men to proprietors, and subjected everyone to consequent disease.

Where other medical texts avoided the subject of passion altogether, Nichols foregrounded the topic in his treatment of the "function of generation." Competing health manuals generally supplied only brief references to conception, usually written in a passive voice and narrated from the point at which the sperm is introduced to the ovum. In contrast, *Esoteric Anthropology* devoted many paragraphs to the encounters preceding zoogenesis. "The expressions of love antecedent to, and connected with its ultimation, are varied and beautiful, involving the whole being," wrote Nichols.

[A] certain warmth and voluptuousness presides over the movements of the body; blushes come often to the cheeks, and the eyes are cast down with consciousness . . . every touch, even of the hem of the garment, is a deep pleasure; the hands clasp each other with a thrill of delight; the lips cling together . . . the bolder hands of man wander over the ravishing beauties of woman; he clasps her waist, he presses her soft bosom, and in a tumult of delirious ecstasy, each finds the central point of attraction and of pleasure, which increases until it is completed in the sexual orgasm—the most exquisite enjoyment of which the human senses are capable. (152–53)

Nichols believed that health necessitated the balanced expression of diverse passions. His vivid writing sought to increase readers' comfort with sexual expression—to own their own desires without shame. "We are not merely to use our eyes and ears, and our legs and arms, but all our organs. Not merely to use them in certain ways, but in all the ways for which we have any natural aptitude or attraction" (253). Denial, implied Nichols, not happiness, was the truly ignoble indulgence. Prim society had it all wrong. *Esoteric Anthropology* asked its readers to imagine respectable men and women (like *themselves*) asserting their needs for sexual happiness.

Grounding all his reform initiatives in medical case studies, Nichols was equally graphic in describing the consequences of sexual disorders, which he believed to derive from ungodly social conditions. "There are thousands of women . . . who never experience the ecstasy of a sexual orgasm," he wrote (198). "There are others in whom it can only be excited with great difficulty, and by various artifices. The more spontaneous the feeling, the less exhausting; the more difficult to excite, the more it tasks the vital energies" (198). This passage not only reiterated the healthiness of women's sexual pleasure, it emphasized that such expression should come naturally and not mechanically—that it should reflect a true and liberated feeling. "I have no doubt, that in a healthy condition, the pleasure of the female is longer continued, more frequently repeated, and more exquisite than that of the male," he continued (153). Yet the line between health and disease remained, and it was strongly determined by personal action. "Men are naturally desirous that their partners should experience pleasure, as it adds to their own," explained Nichols (198). "To effect this, they resort to manipulations of the clitoris, with their fingers, etc., and to various novel, and, to a certain extent, unnatural methods and positions. Out of these grow terrible mischiefs, especially to women. Many of the worst cases of ovarian and uterine disease are caused by the forced pleasures of artificial excitement" (198).

Clearly, Nichols believed that the biological and political aspects of sexual intercourse demanded forthright discussion. It was the execution of this belief that set *Esoteric Anthropology* outside the mainstream of domestic health manuals. Regarding women's physical self-ownership and sexual freedom, Nichols pleaded with his readers: "O Woman! You must accept this responsibility, and you must demand and have these rights. When men are once enlightened on this subject, none but inhuman wretches and monsters will deny them" (235).

In addition to its pervasive emphasis on women's rights, *Esoteric Anthropology* also included discussion of topics either absent from or only briefly alluded to in other works on domestic medicine. A chapter on "Passional Diseases" addresses homesickness, extreme religious fervor, undue "acquisitiveness," jealousy, and "love sickness," as spiritual disorders—"morbid affections of the faculties and passions of the soul" (329). Treatment for these problems, Nichols explained, was analogous to that for organic disease: the "development and harmonization" of one's faculties and passions, which entailed not only a careful, vegetarian diet, regular outdoor exercise, and daily bathing, but also an effort to "energize the mind, and purify it of false ideas" (332–33). To this end, Nichols recommended the writings of Stephen Pearl Andrews and Charles Fourier—both nonmedical social reformers who sought utopian, communal solutions to the misery, boredom, and ill health evidenced by thousands.[10]

Other "disorders" that received respectful attention in *Esoteric Anthropology* included homosexuality, "bestiality," and incest. While characterizing homosexuality as "unnatural" ("filthy" in the case of men and "false" in the case of women), Nichols emphasized that same-sex intercourse "has been practiced from the remotest ages," in a variety of social settings, and in many countries. "I see no reason for punishing a man for an act which begins and ends with himself, or with a consenting party," he asserted in reference to homosexuality—a view still considered radically liberal by many Americans in the twenty-first century (201). As for intercourse with animals, Nichols wrote only: "I should think the act was punishment enough. If not, exposure and the attending disgrace would be" (202). Yet incest, he believed, was "a very different matter," as it risked "scrofulous, insane, or idiotic offspring"—a physiologic argument, rather than a moral one (202). Though not elaborate, the mere acknowledgment of these subjects was as daring as his claims that women given over to excessive sexual indulgence might experience "six or seven orgasms in rapid succession, each seeming to be more violent and ecstatic than the last . . . accompanied with screams, bitings, [and] spasms" (200).

Although these passages were bold and relatively shocking, most of Nichols's 482 pages did not challenge traditional medical knowledge, undermine accepted social mores, or drag gentle readers into unexpectedly rugged terrain. The artful manner in which Nichols packaged his radicalism deserves attention, for *Esoteric Anthropology* surely reached a wider and more diverse audience than did many fringe pamphlets on free love.[11] Though the book was certainly an outlier, recall that the latest edition appeared sixty-three years after the first; its appeal surely owed much to its quality as a thought-provoking health manual. Most of what Nichols taught corroborated the technical, scientific wisdom of allopathic physicians and validated the hygienic advice that pervaded the popular health movement generally. Thus even as Nichols drew readers into contested political territory, he constantly provided reassurance of his legitimate medical expertise and reinforced what readers already believed to be intelligent personal choices.

Even when echoing the sentiments of other medical writers, however, Nichols likely kept his readers awake with the explicit nature of his prose. Writing, for example, on the subject of masturbation (which he, like other contemporary medical authorities, characterized as "truly a matter of life and death" [375]), Nichols explained: "A child of a year old may get into this habit, either from some perverted instinct or from some irritation of the genital organs. This is especially the case with little girls. There is some itching of or around the clitoris, the child begins to scratch or rub the parts, and is astonished to find the friction producing keen sensations of pleasure. It is repeated next day with the same result—the habit is formed, and the nervous system is wrecked" (375).

As much as 90 percent of the general population engaged in this "self pollution," estimated Nichols—a statistic that probably comforted many readers (201). A person too ashamed to consult personally with a physician would find unambiguous descriptions of habits and symptoms in Nichols's work, as well as forthright medical advice. *Esoteric Anthropology* offered readers a less squeamish—and more circumstantial—discussion of important health issues than did many of its competitors. Its vivid descriptions of physical experiences alternated with presumptuous (and no doubt entertaining) claims about those who had them: "[W]omen who have exhausted themselves by secret licentiousness [masturbation] are often so *virtuous* as to hate the sight of a man, and abhor the idea of the holiest expression of the Divine Creative passion.

These are our most censorious prudes, and immaculate virgins, who do not fail to crush and banish from their pure society, any poor girl who yields to the supplications of her lover and her own natural healthy desires" (402). How have such women destroyed themselves, and what becomes of them? "[I]f . . . the vagina itself is the seat of abuse, the artificial instruments made use of, and even their increasing size, destroy the proper effect of the natural organ. When woman so unfortunate comes to be married, she receives the warm embraces of her husband with indifference, and perhaps with disgust or absolute pain. She is cold amid his ecstasies, yields only to his commands, and turns from him with repugnance" (403).

Despite his journalistic flair for graphic imagery, Thomas Low Nichols also produced sensible, familiar medical prose: "Prevention here is the all-important thing. Every man and woman should endeavor to have such a healthy control . . . as to avoid giving their children the terrible inheritance of diseased and disordered passions . . . [T]here is no violation of nature which brings not its penalty" (403).

The notion that children suffered constitutionally from birth as a result of parental disease held wide currency in the nineteenth century. Hence many medical advisers emphasized the critical importance of maternal well-being, especially during pregnancy and while nursing children. Almost universally, women were counseled to avoid sexual intercourse when pregnant or nursing. As a physician, Nichols strongly endorsed such constraint, writing that no nursing woman "should ever be exposed to sexual excitement," as such exposure would reduce both the quantity and quality of her milk (238). The most common cause of abortion, he taught, was intercourse during pregnancy. "All amative excitement . . . perils the existence, as it injures the proper growth, and injuriously affects the character of the child," wrote Nichols (191).[12] For the sake of vulnerable offspring as much as for the sake of the mother, he reiterated the need for female control of sexual frequency. For a man to force intercourse "at any time, and especially during pregnancy, can not be called beastly," he wrote, "for it would be a libel on the brutes" (152).[13]

Nichols's commitment to women's health derived in large measure from the powerful influence of his wife, Mary S. Gove Nichols, who in 1838 had become the first woman in America to offer public "lectures to ladies" on the subjects of anatomy, physiology, and hygiene. Between 1847, when they met, and 1853, when Nichols published *Esoteric Anthropology,* the couple had become prominent leaders of the water cure movement, each having published independent monographs on its methods as well as numerous brief articles. Already a successful hydropathic practitioner when she met her husband, Mary Gove Nichols had used her own earnings to support his allopathic medical education at the University of the City of New York; she had also trained him in the methods of water cure. Together, they contributed regularly to the thriving *Water-Cure Journal, and Herald of Reforms*, edited by their colleague, Russell T. Trall.

The survivor of an abusive first marriage during which she suffered four stillbirths, Mary Gove Nichols unquestionably helped to shape her husband's views on the physiological importance of women's sexual and emotional happiness. Had he been a less enlightened man when they met, however, Mary Gove would never have married Thomas Nichols. In her autobiography, Mary Nichols described the firm conditions she had established for their union: "In a marriage with you, I resign no right of my soul. I enter into no compact to be faithful to you. I only promise to be faithful to the deepest love of my heart. If that love is yours, it will bear fruit for you . . . If my love leads me from you, I must go . . . I must keep my name—the name I have made for myself, through labor and suffering . . . I must have my room, into which none can come, but because I wish it."[14] The couple sought to live what they preached and, through their writing, to convert the world.

Soon after the birth of their only child, the Nicholses founded the country's first hydropathic medical school, the coeducational American Hydropathic Institute. There, they trained other water cure practitioners. It was in this setting that Thomas and Mary Nichols developed the curriculum that would serve as armature to *Esoteric Anthropology*. When he finally published the work in 1853, Nichols joyously expressed the hope that his text might "be one of the instrumentalities of a real, physical redemption for mankind," out of which would "be developed all moral excellence, intellectual elevation, social harmony, and individual and general happiness" (7–8). To his dismay, the *Water-Cure Journal* refused to acknowledge the book's existence.

Revolted by its free-love perspective as well as its sexually graphic content, Russell Trall blocked all reference to the text in his prominent journal, barring even paid advertisements. Though he had been a regular contributor to the *Water-Cure Journal* for years, Nichols was forced to promote his book elsewhere. The bitterness engendered by this lack of support led to the Nicholses' departure from the *Journal* altogether. As Thomas Nichols described the experience, "It was like being turned out of a railway-train, in which our friends had taken passage."[15] Their relationships in New York grew so strained that soon the Nicholses relocated to Yellow Springs, Ohio, purchasing land adjacent to the new and relatively liberal Antioch College. There too, Thomas and Mary Nichols met with hostility. On Antioch's campus, a student was expelled for distributing works by the infamous Nicholses. It was all the more shocking to radicals and conservatives alike when, in 1857, Mary and Thomas Nichols announced their conversion to—of all things—Catholicism.[16] Between 1857 and 1861, when the Nicholses moved permanently to England, the couple and their young daughter traveled to a number of Catholic institutions throughout the United States, where Thomas and Mary offered lectures on health and hygiene.

Later editions of *Esoteric Anthropology* reflect the changes in Nichols's belief system.[17] No longer a free-lover but rather a proclaimed Catholic, he took care to eliminate passages that contradicted church teachings. For example, he deleted a passage from the 1853 edition characterizing celibacy as "a violation of nature" comparable to abortion: "[T]hey have each the same social result, with respect to population. I can not decide as to their relative badness" (193). In 1853, he had argued against legal interference with a woman's right to abortion, asserting that before birth, the mother's fertilized ovum is "as yet only an organic structure, taking its sustenance from her. It is an unnatural thing for her to refuse this sustenance—it may be very wicked. But it is exclusively her own affair. The mother, and she alone, has the right to decide whether she will continue the being of the child she has begun" (190). In the corresponding passage of his 1870s edition, Nichols wrote: "[W]hen a woman has united with a man in the creative act, the life of the being so formed is sacred. From the moment of conception it is a human life with all its possibilities, temporal and eternal" (136). Despite this new perspective, however, Nichols still believed in the universal right to knowledge. On the very same page that he declared a fetal life sacred, Nichols introduced a discussion of the various means of procuring abortion—a passage little altered from the 1853 original—and one that includes a description of the surgical procedure most safely employed: place a "slender bougie into the uterus and let it remain until it induces contractions" (138).

Rewriting his book, Nichols also toned down its ecstatic descriptions of sexual orgasm. Perhaps they seemed excessive to his more mature eye. Other views changed as well. In 1853, Nichols had argued that the persistence of sexual desire after menopause, along with the "delicious pleasure [sexual union] brings, in a healthy state, to both sexes, all point to other uses and ends than those of procreation" (222). He felt strongly about the subject: "The woman who has borne children through the menstruating period, has now a compensation in the full pleasure of love, without its privations or cares . . . Nature has been very bountiful in the distribution of the sources of happiness; it is man alone that is niggard and perverse" (223). His revised work from the 1870s, however, explained that "we should detest any animal that made pleasure alone the motive of sexual union" and that there "can be no doubt that this is also the natural and healthful law of the human species" (117). The revision is consistent with Nichols's recantation of free love, which is reflected in other editorial changes as well. He omitted passages exploring such questions as "Can One Love Two Or More Persons At Once?" and "Is Love Enduring?" Indeed, he seemed determined to recreate himself. It is difficult to believe that the following passage from the 1870s edition was written by the same Thomas Nichols who in 1853 had uncritically described contraceptive methods, including the condom, the sponge, the cold-water douche, compression of the penis during ejaculation, and withdrawal: "The secularist philanthropists who . . . [advise] . . . frequent and regular exercise of amativeness; who hold that what good men in all ages have called virtue is a vice, that chastity is wickedness and continence criminality, and that lewdness, fornication, and adultery are moral duties, are obliged also to advocate the use of preventive checks to an increase of population . . . The very thought and intention of enjoying a natural pleasure and at the same time doing something to hinder the natural effect of such enjoyment, is a source of evil" (114–15).[18]

Despite his new conservatism, however, in the 1870s edition Nichols did not waver in his belief that if contraception "can ever be justified, it is when a woman is unwillingly compelled to submit to the embraces of her husband" (114). The relationship between Nichols's Catholicism and his devotion to women's rights is difficult to untangle.

Clearly, Nichols ceased to proclaim the universal right to free divorce. Yet unlike his explicit reversal regarding the status of the unborn fetus, his later opinions regarding divorce quietly preserve aspects of his earlier political convictions. In 1853, he had written: "Entire freedom of divorce could never dissolve one real marriage; and if the shams were broken, real ones could take their places. Many of the happiest marriages I know are between those who have been unhappily married and separated. Because a man makes an innocent blunder, it is no reason he should suffer for it a life time" (218). His 1870s revision simply asserts, "The Catholic Church teaches that marriage is a sacrament, and therefore it does not permit divorce, but only separation; neither party being allowed to marry again until the death of the other" (96). While his original work did not reference the views of the Catholic Church, his revision did not reference his own views on this point. Elsewhere, however, he wrote:

> If mutual love be the sole justification of sexual union—if it be false, unnatural, abhorrent, where such love does not exist, then the cessation of love on the part of either would be the end of marriage—a divorce, or at least a separation. But the interests of children, families, and society, as it is now constituted, do not permit of divorce for sentimental grievances. The intimate relations of two married partners must be regulated by themselves . . . Where there is no positive sin, no violation of conscience, we must seek the greatest good, even in a choice of evils; and we are not to seek our own good merely, but to make sacrifices, if need be, for the good of others.[19]

Here, beneath his conservative words, Nichols implicitly seems to tolerate a window for divorce when there *has* been either "positive sin" or "violation of conscience." Catholicism had genuinely replaced his youthful liberalism, but it had not undermined his sympathy for victims of spousal abuse.

Thomas Low Nichols was intriguing. His early attachments to the free-love movement and later to the Catholic Church set him well apart from fellow laborers in the field of popular health reform. Despite these differences, however, he produced works of broad social appeal. In many ways, even the first edition of *Esoteric Anthropology* was a representative health manual. It taught readers the bones of the skull, the functions of the skin, and the distribution of the nerves. It described the nature of smallpox and the history and symptoms of syphilis. It served as a manual of midwifery and a guide to healthy eating. It equated the laws of nature with the laws of God. The work's special qualities lay in its determination to interpret those laws afresh—to wield physiology as a tool for radical social change. As Nichols described his first edition in closing: "I have not written a book on morals, but on science, which is the true basis of morality. But is must be evident that in a discordant society, as in an unhealthy individual, moralities grow morbid. Some of the most crushing sins against social laws, are acts of the simplest conformity to natural law" (477).

Lest the implications of this logic go unnoticed, Nichols, as always, undergirded his claim with the most forthright example: "What is more natural than that a healthful, passionate woman should give herself in love to a man whom she believes to be worthy of her? The stronger, and healthier, and better she is, the more she is impelled to such an act . . . I feel bound in honor to say, that, so far as my observation extends, the women most likely to outrage society, in

love relations, are the truest, the noblest, the greatest, and those we should
most delight to honor" (477–78).

In fact, Nichols *had* written a book on morals. He rebuked society for its
damaging behavioral expectations and he challenged readers to adopt new
ways. Contemptuous of sexual double standards, he maintained in the 1870s
edition, "What is right for one sex must be right for both; what is wrong for one
must be wrong for both" (117). Thus Thomas Low Nichols left historians a fascinating and important document. Through *Esoteric Anthropology*, we may
survey the limits of antebellum medical knowledge; we may glimpse patients'
most embarrassing symptoms; and we may eavesdrop on the emotional concerns they brought to their physicians. We may also gain greater understanding of mid–nineteenth-century radicalism, which clearly relied on a varied
print media to foster intellectual and political change.

Popular health literature of the era reflects heated competition among diverse medical sects; personal financial needs motivated many authors. But, as
is obvious from *Esoteric Anthropology*, far more was at stake in these texts than
sales and advertising. The genre of domestic medical texts deserves attention
as a political force, for its agenda extended well beyond a prudent response to
the common cold. With rapid advances in publishing technologies, transportation, and general rates of literacy, religious writings lost their unchallenged dominance as printed sources of moral guidance. Medical texts gained
an authoritative foothold and began to introduce philosophies often radically
different from those of routine pulpit sermons. The health reform movement
reflected an enormous interest in the ideas of new experts, who also spoke
knowingly of God's laws. As Nichols wrote in 1853: "No *natural* passion, no
healthy attraction of any being is wrong; for that Being, God, who distributes
attractions, governs the harmonies of the universe. We produce only discord,
when we use our freedom to oppose them" (212).

Perhaps most strikingly, *Esoteric Anthropology* illuminates the force with
which men could and did speak publicly on behalf of women's rights. The feminism embedded in many alternative medical writings was broad and modern
in its demands. As another leading water cure physician had put it in 1852:
"[Woman] should make it absolutely impossible for man to philosophize correctly, to exhort edifyingly, to enter into society properly, to legislate with per-
manency, unless he proceeds on the recognition that woman, as well as man, has
a soul, an organization, an entity not to be buried up in him, or in arrangements
solely contemplating him. I would have them do this, because the differences

between woman and man are *vital*—difference which God established . . . and without which the ends of existence cannot be answered."[20]

The logic of feminist health texts began with the accepted notion that women were inherent caretakers—the natural managers of household health. This assumption was especially influential for advocates of women's medical education, who considered women innately suited to the physician's role. Rather than narrow his cause to a particular area of social equity such as professional training or the right to vote, however, Nichols chose to cast his net widely. Defining happiness as *prerequisite* to health—and vice versa—Nichols argued for the social conditions he believed necessary for universal happiness. These conditions began with the sexual and economic liberation of women.

Although a bolder writer than most of his contemporaries and more explicitly devoted to women's sexual self-ownership and freedom than other health reformers, Nichols was not a marginal figure. His medical writings echoed those of his peers. He reached a wide audience. The success of *Esoteric Anthropology* reveals a quiet, grass-roots mobilization. The speeches of suffragists clearly fell on the ears of many who had already adopted personal habits authoritatively claimed to foster greater empowerment—people who might not support so public a goal as women's suffrage but who had in fact voted for change by choosing to read books like *Esoteric Anthropology*. In the 1853 edition, Nichols had framed the woman problem physiologically: "I can not undertake to reconcile the teachings of nature with the laws of society. They are everywhere at variance, and all our miseries come out of these discords" (151). Nichols did not blame readers personally for their ailments. Rather, he condemned what he perceived as dangerous, misguided social expectations placed on Americans—expectations indoctrinated from birth and understandably difficult to overcome. He spoke to his readers collectively as intelligent victims, seeking with *Esoteric Anthropology* to empower once again "all those wise enough to profit by its teachings" (5). When a reader put down *Esoteric Anthropology*, she or he had been encouraged not only to bathe regularly, eat moderately, dress comfortably, and exercise well, but also to question authority. Why should readers sustain arbitrary institutions that repressed, denied, suffocated, or saddened? For as Dr. Thomas Low Nichols believed, "We should give ourselves all the advantages of a full, healthy, integral life; a life of energy, activity, beauty, and enjoyment" (80).

NOTES

1. Thomas Low Nichols, *Esoteric Anthropology. A Comprehensive and Confidential Treatise On The Structure, Functions, Passional Attractions And Perversions, True And False Physical And Social Conditions, And The Most Intimate Relations Of Men And Women* (New York: Stringer & Townsend, 1853), 5, italics in original. Subsequent references to this work appear parenthetically in the text (they are the only references treated this way in the chapter).

2. Thomas Low Nichols, *Nichols' Health Manual: Being Also A Memorial Of The Life And Work of Mrs. Mary S. Gove Nichols* (London: The Author, 1886), 91.

3. Other prominent health writers addressing such subjects included William Andrus Alcott and Frederick Hollick, both of whom were widely read. For elaboration on the genre of popular health texts and on the career of Frederick Hollick, see chapter 1 of this book; for more information on nineteenth-century medical turf wars, see Alex Berman, *America's Botanico-Medical Movements: Vox Populi* (New York: Pharmaceutical Products Press, 2000); Martin Kaufman, *Homeopathy in America: The Rise and Fall of a Medical Heresy* (Baltimore: Johns Hopkins University Press, 1971); idem, *American Medical Education: The Formative Years, 1765–1910* (Westport, Conn.: Greenwood Press, 1976); Joseph F. Kett, *The Formation of the American Medical Profession: The Role of Institutions, 1780–1860* (Westport, Conn.: Greenwood Press, 1980); and Paul Starr, *The Social Transformation of American Medicine* (New York: Basic Books, 1982). On the water cure movement, see Susan Cayleff, *Wash and Be Healed: The Water-Cure Movement and Women's Health* (Philadelphia: Temple University Press, 1987); and Jane B. Donegan, *"Hydropathic Highway to Health": Women and Water-Cure in Antebellum America* (New York: Greenwood Press, 1986).

4. Thomas Low Nichols, *Woman, in All Ages and Nations; A Complete and Authentic History of the Manners and Customs, Character and Condition of the Female Sex, in Civilized and Savage Countries, From the Earliest Ages to the Present Time* (New York: H. Long & Brother, 1849), 229; and idem, "Vincent Preissnitz [*sic*]," *Nichols' Journal—A Weekly Newspaper, Devoted To Health, Intelligence, Freedom; Individual Sovereignty And Social Harmony* 2, no. 1 (1854): 4.

5. Self-publishing was a common and respectable practice in the mid–nineteenth century, free of the stigma associated with today's vanity presses. Most publishing houses happily issued books over their regular imprint that were printed at the risk and expense of the author. The publisher billed the author for printing and binding and charged a commission for distribution. Thus a new writer could enjoy the prestige of seeing his or her work carried by a major publishing house, though most publishers did not push such books and usually the authors lost money on the venture.

Nichols self-published *Esoteric Anthropology* somewhat differently. Perhaps because he knew a regular publisher would be unlikely to promote the work, he assumed not only the expense of manufacture and the risk of loss but also the labor of distribution. He purposely kept it "out of the trade" and relied on direct mail service to deliver individually wrapped copies, per order. He also recruited agents and peddlers with "buy ten get one free" offers, known to contemporaries as subscription clubs. That Nichols decided to contract with Stringer & Townsend suggests that the flow of orders had become unmanageable. The post office would not ship large crates of books, and ordinary freight usually had to be transferred from one rail or steam line to another at each leg of the journey, often resulting in long delays. Express companies were beginning to solve this problem, but at high cost. Regular publishing houses, with established networks, managed bulk distribution more easily. Stringer and Townsend were in business together between 1848 and 1857. In 1848, Stringer left Burgess, Stringer & Co. to found Stringer & Townsend, which went on to publish two dozen books by James Fenimore Cooper, as well as many popular, inexpensive works.

It seems that Nichols continued to self-publish even after he contracted with a regular publisher—also a fairly common practice. This allowed authors to retain greater control over their work and to keep more of the profits. Control over a particular text could be achieved by having it stereotyped and then retaining ownership of the thin metal plates, leasing or loaning them to a publisher. Since Nichols already owned plates for his book, he was in a strong position to negotiate favorable terms with his publishers. These plates stayed with Nichols through his life and enabled him to make revisions in subsequent editions of the book.

James Green, Associate Librarian, Library Company of Philadelphia, telephone conversation with author, 12 July 1999. See also Michael Winship, *American Literary Publishing in the Mid–Nineteenth Century: The Business of Ticknor and Fields* (Cambridge: Cambridge University Press, 1995); and William Charvat, *Literary Publishing in America, 1790–1850* (Amherst: University of Massachusetts Press, 1993). It should be noted that estimates given of the sales of *Esoteric Anthropology* come from Nichols's own reports. See Nichols, "Esoteric Anthropology," *Nichols' Journal of Health, Water-Cure, and Human Progress* 1, no. 3 (1853): 21; Nichols, "American Hydropathic Institute," *New York Daily Tribune*, 22 July 1853, 7; and idem, *Nichols' Monthly Extra* (ca. 1856), 4.

6. Nichols, "Esoteric Anthropology," *Nichols' Journal of Health, Water-Cure, and Human Progress* 1, no. 1 (1853): 8. Karen Lystra has documented the growing significance of "private space" for Americans during the nineteenth century; her study of love letters argues that despite their popularity, restrictive advice manuals regarding social comportment did not hinder private expressions of passion. Nichols's emphasis on the private nature of his work implicitly invited an honest and free emotional response to its content. One of his chief goals was the liberation of socially repressed passion. See Karen Lystra, *Searching the Heart: Women, Men, and Romantic Love in Nineteenth-Century America* (New York: Oxford University Press, 1989), 12–27.

7. Mary Gove Nichols, "Esoteric Anthropology," *Nichols' Journal of Health, Water-Cure, and Human Progress* 1, no. 1 (1853): 8.

8. Nichols and Nichols, *Marriage: Its History, Character, and Results; Its Sanctities, And Its Profanities; Its Science And Its Facts. Demonstrating Its Influence, As A Civilized Institution, On The Happiness Of The Individual, And The Progress Of The Race* (New York: The Authors, 1854).

9. Ibid., 89, 265.

10. The social theories of Charles Fourier attracted considerable attention in the United States when the *New York Tribune* began in 1842 to carry columns by Fourier disciple Albert Brisbane. Thousands were drawn to the concept of "association," and many established experimental living arrangements based on Fourier's blueprints for group living. See Carl Guarneri, *The Utopian Alternative: Fourierism in Nineteenth-Century America* (Ithaca, N.Y.: Cornell University Press, 1991); and Frank E. Manuel and Fritzie P. Manuel, *Utopian Thought in the Western World* (Cambridge: Harvard University Press,

1979). On the fascinating life of Stephen Pearl Andrews, see Madeline B. Stern, *The Pantarch: A Biography of Stephen Pearl Andrews* (Austin: University of Texas Press, 1968).

11. For an example of Nichols's own contribution to this genre, see Thomas Low Nichols, *Free Love: A Doctrine of Spiritualism. A Discourse Delivered In Foster Hall, Cincinnati, December 22, 1855* (Cincinnati: F. Bly, 1856). This rare pamphlet is held in the Western Reserve Historical Society, Cleveland, Ohio.

12. Fellow water cure advocate Russell T. Trall concurred that "excessive sexual indulgence" was a "frequent cause" of abortion. R. T. Trall, *The Hydropathic Encyclopedia: a System of Hydropathy and Hygiene* (New York: Fowlers and Wells, 1853), 2:458. At an earlier time, it was believed that "Venery" during the first two months of pregnancy could predispose to "a Mole or Superfaetation; which is the adding of one *Embrio* [*sic*] to another." *Aristotle's Compleat and Experienc'd Midwife* (London: William Salmon, 1721), 29–30.

13. Similar views regarding the dangerousness of parental lust to gestating offspring can be found throughout nineteenth-century health-reform literature. As John Harvey Kellogg later put it, "If a child is begotten in lust, its lower passions will as certainly be abnormally developed as peas will produce peas." John H. Kellogg, *Plain Facts About Sexual Life* (Battle Creek, Mich.: Office of the Health Reformer, 1877), 64–69. See also John Cowan's *The Science of a New Life* (New York: Cowan & Co., 1880), 210–14, where it is argued that parents who "desire a child that will be the embodiment of licentiousness" need only "put to shame the beasts of the field in their unnatural lust" during the period of gestation. The Cowan and Kellogg quotes are cited in Ronald G. Walters, *Primers for Prudery: Sexual Advice to Victorian America* (Englewood Cliffs, N.J.: Prentice-Hall, 1974), 52, 152–53.

14. Mary Gove Nichols, *Mary Lyndon: or, Revelations of a Life. An Autobiography* (New York: Stringer & Townsend, 1855), 385.

15. Nichols, "The Reason Why," *Nichols' Journal* 1, no. 1 (1853): 5. Trall had also probably taken offense at several of the anatomical engravings included in *Esoteric Anthropology*, for example, images of erect penises.

16. The story of this dramatic conversion extends beyond the scope of this chapter. For more information on the Nicholses and the water cure movement, see Jean Silver-Isenstadt, *Shameless: The Visionary Life of Mary Gove Nichols* (Baltimore: Johns Hopkins University Press, 2002). For detail regarding the Antioch student, see this pamphlet: Jared Gage, *Address To The Friends, Officers, And Students of Antioch College*, ca. 1856, Antioch College Library Archives.

17. A reprint of an 1870s edition of *Esoteric Anthropology* was published in 1972 by Arno Press in its series Medicine and Society in America, which was edited by Charles E. Rosenberg.

18. In his *Human Physiology. The Basis of Sanitary and Social Science* (London: Trubner & Co., 1872), Nichols explicitly argues that sexual intercourse is justified only for reproduction—obviously a substantial transformation of belief.

19. Nichols, *Human Physiology*, 309–10. In other writings, Nichols praised Catholics for their unwillingness to get involved in state politics. He lacked respect for any priest who sought to influence the votes of his parishioners. Therefore, it may be fair to interpret the revised passage from *Esoteric Anthropology* as endorsing church views *for Catholics*, but not necessarily for society at large. Nichols valued the separation of church and state. See Thomas Low Nichols, *Forty Years of American Life* (London: J. Maxwell, 1864; New York: Negro Universities Press, 1968), 1:375.

20. James Caleb Jackson, *Hints on the Reproductive Organs: Their Diseases, Causes, and Cure on Hydropathic Principles* (New York: Fowlers & Wells, 1852), 30. Jackson hoped that women's "distinctiveness" would "be so broad as to secure for them personality, identity, *self*-ownership." He went on to say: "In the reorganization of society on the basis of free womanhood, or the right of woman to herself, in my judgement it would be of little consequence whether *her* sphere of activity and man's prove to be identical. The sphere of woman is yet to be defined" (30–31, italics in original).

9

RONALD L. NUMBERS

Sex, Science, and Salvation
The Sexual Advice of Ellen G. White
and John Harvey Kellogg

The flamboyant Dr. John Harvey Kellogg—inventor of flaked cereals and peanut butter, promoter of frequent bowel movements, author of best-selling sexual and dietary manuals, and prolific sanitarium builder—is no stranger to students of American medicine and culture. A sampler of scholarly and popular opinion corroborates this claim. James C. Whorton describes him as "the most formidable reformer of American living habits of the twentieth century." William R. Hunt says he was "probably the most famous physician in America" during the early years of the century. Peter Gardella credits him and his church with fomenting "a revolution in medical thought" about sexual pleasure. John Money asserts that, "because of his medical status and his immense popular prestige, Kellogg was more responsible than any other person of his generation for popularizing the fallacious disease of masturbation." "For sheer professional arrogance and anachronistic disregard of scientific medical knowledge," writes the Johns Hopkins sexologist, Kellogg's *Plain Facts about Sexual Life* was "unsurpassed in medical nonsense." With an estimated half-million copies in print by the early twentieth century, it was also unsurpassed in circulation.[1]

As Money's appraisal suggests, by the latter twentieth century Kellogg's sex books had become some of the most reviled symbols of everything wrongheaded about the sexual advice of an earlier age. In 1964 the sexual revolutionary Hugh M. Hefner devoted a sixteen-page installment of "The Playboy

Philosophy" to the horrors found in Kellogg's "manual of love and marriage," and in 1993 the novelist T. Coraghessan Boyle featured the doctor's obsession with sex and feces in *The Road to Wellville*. In the movie version of this book, the Oscar-winning actor Anthony Hopkins immortalized Kellogg as a buck-toothed huckster who discovered in health "the open sesame to the sucker's purse."[2]

Kellogg scholarship has tended to fall into two camps. Historians of sexual advice and health reform have been inclined, not without warrant, to see Kellogg as a late-nineteenth-century disciple of Sylvester Graham, William Alcott, and other postmillennialist health reformers who optimistically proclaimed a gospel of health to usher in the millennium preceding Christ's return to earth.[3] They have paid virtually no attention to the influence of the premillennialist Seventh-day Adventist prophetess Ellen G. White, Kellogg's foster mother, spiritual adviser, and instructor in health reform, who awaited the imminent end of the world. Biographers of Kellogg, in contrast, have largely ignored his writings about sex while emphasizing the formative role played by White.[4] Because none of the former seem ever to have looked at White's two booklets on sex, *Appeal to Mothers* (1864) and *Solemn Appeal* (1870), they have largely overlooked how indebted Kellogg was to White; because the latter have neglected Kellogg's sex-oriented writings, they, too, have missed the connection. My primary goal is to explain how Kellogg succeeded in translating White's sexual warnings, born of visions and bred in a culture of religious and medical sectarianism, into arguably the most widely circulated medical text on sex in the years spanning the last two decades of the nineteenth century and first two decades of the twentieth century.

The broad contours of the history of American sexual-advice literature are well known. As Stephen Nissenbaum has pointed out, Sylvester Graham's pioneering *A Lecture to Young Men on Chastity* . . . (1834) broke with the older moralistic literature on the subject in two ways: It was based largely on scientific rather than biblical arguments, and it focused not on the sins of adultery and fornication but on the previously neglected problems of masturbation and marital excess, which Graham defined for most people as intercourse more often than once a month. Graham, a Presbyterian minister and temperance lecturer, emphasized the intimate connection between diet and sex resulting from the arousing effects of stimulating foods on the sexual passions. He recommended, as one of the best means of controlling unwholesome urges, the adoption of a meatless diet and the forsaking of condiments, spices, alcohol, tea, and coffee.[5] From the time that Graham's book appeared, the subject of sex played a highly visible role in health-reform literature. William A. Alcott, Mary Gove, Samuel Gregory, Russell T. Trall, and James Caleb Jackson all discussed the dangers of what they regarded as excessive or abnormal sexual activities, particularly masturbation, which they believed caused a frightening array of pathological conditions ranging from dyspepsia and consumption to insanity and loss of spirituality. In the mid-1830s Samuel B. Woodward, the superintendent of the Worcester Lunatic Asylum in Massachusetts, began drawing attention to the growing number of young people made mad by masturbation. By carefully couching their appeal in humanitarian terms, such writers sought to avoid offending the sensibilities of a prudish public. Together they launched a moral crusade against what Jackson called *"the great, crying sin of our time."*[6]

Visions of Sex

The roots of John Harvey Kellogg's philosophy of diet and sex ran directly back to Ellen G. White, the Seventh-day Adventist prophetess who turned health reform into a sacred duty. A self-described "lifelong invalid" with little formal education, White early on became obsessed with religion, especially after she and her family embraced William Miller's predictions about the end of the world in the early 1840s. Shortly after the Millerites' Great Disappointment on October 22, 1844, when Christ failed to return to earth, White, then aged seventeen, began having dissociative experiences or "visions," during which heavenly beings would show her events past, present, and future. Her vision-based "testimonies" served, in conjunction with the Bible, as the intellectual foundation and behavioral guide for her followers, who by the early 1860s numbered about 3,500 and were calling themselves Seventh-day Adventists.[7]

White, somewhat disingenuously, liked to date her discovery of health reform to a vision in June 1863, during which God showed her the evils of medicinal drugs, alcohol, tobacco, tea, coffee, meat, spices, fashionable dress, and sex and the benefits of a twice-a-day vegetarian diet, generous use of water for drinking and bathing, fresh air, exercise, and a generally abstemious lifestyle. Actually, she had known about the health-reform movement for some time and had recently used the water treatments recommended by some of its leaders to save the lives of two of her sons. Besides, her eldest son, Edson, had turned thirteen in 1863 and had begun to display some disturbing behaviors: a lack of interest in religion, a "passion" for reading storybooks, a fondness for girls, and irresponsibility—all characteristics she, and many of the health reformers, associated with self-abuse or masturbation. By 1864 Ellen White had acquired at least Trall's and Jackson's books on sex and was penning her first work on health, a plain-looking pamphlet reminiscent in content and appearance of Mary Gove's *Solitary Vice: An Address to Parents* (1839), printed in White's hometown of Portland, Maine. White entitled her work—published by the Seventh-day Adventist Publishing Association that her husband, James, had founded in Battle Creek, Michigan—*An Appeal to Mothers: The Great Cause of the Physical, Mental, and Moral Ruin of Many of the Children of Our Time* (1864).[8]

Writing for an Adventist audience expecting the imminent end of the world, White warned that "solitary vice" would ruin life and health on earth and preclude a future existence in heaven. She told of how her unnamed angel guide had exposed her to the horrors of human depravity. "Everywhere I looked," she recalled as though describing a real event, "I saw imbecility, dwarfed forms, crippled limbs, misshapen heads, and deformity of every description"—all the result of the practice of solitary vice, so widespread that "a large share of the youth now living are worthless." Even adults had fallen victim to this satanic lure. At one point in her vision she recognized an acquaintance, "a mere wreck of humanity," who had been brought near death by this demonic habit. To drive her message home, White noted that continued masturbation would produce not only hereditary insanity and deformities but a host of diseases, including "affection of the liver and lungs, neuralgia, rheumatism, affection of the spine, diseased kidneys, and cancerous humors." Not infrequently, it led its victims "into an early grave."[9]

Combatting this terrible curse required self-control, an abstemious diet, parental vigilance, and warnings about the loss of eternal life. White thought it important "to teach our children self-control from their very infancy, and learn them the lesson of submitting their wills to us." Like Graham before her, she regarded a bland diet as one of the best means of curbing the urge to masturbate. She proscribed all stimulating substances such as "mince pies, cakes, preserves, and highly-seasoned meats, with gravies," since they created "a feverish condition in the system, and inflame[d] the animal passions." In addition to guarding their children's diets, parents were to watch constantly for signs of self-abuse: absent-mindedness, irritability, forgetfulness, disobedience, ingratitude, impatience, disrespect for parental authority, lack of candor, a strong desire to be with the opposite sex, and a diminished interest in spiritual things. If apprehended in the act, the children were to be told: "[I]ndulgence in this sin will destroy self-respect, and nobleness of character; will ruin health and morals, and its foul stain will blot from the soul true love for God, and the beauty of holiness."[10]

White believed that special care should be taken to protect the young from the contaminating influence of other children. As an adult she had come to view a crippling childhood accident, which had left her an invalid for years, as a blessing in disguise, because it had preserved her innocence. Self-conscious about her intimate knowledge of masturbation, she insisted that she had grown up in "blissful ignorance of the secret vices of the young" and had learned about them only after marriage, from "the private death-bed confessions of some females." Perhaps she protested too much. From her own testimony we know that at about age twelve, near the onset of puberty, she felt terribly guilty, unworthy, and sinful. Could these feelings have arisen from the first stirrings of adolescence or from the sexual exploration of her own body, common to children of her age? To maintain the purity of her own offspring, she never permitted them to associate with "rough, rude boys" or to sleep in the same bed or room with others of their age.[11] She expressed particular concern about two neighbor boys, Samuel and Charles Daigneau, who, she saw in vision, had "gone to great lengths in this crime of self-abuse"—so great that Charles was losing his intellect and eyesight. (Somehow he survived to age seventy-one without going blind or insane. Samuel went on to serve two terms in the Michigan state senate and died at age eighty-two after enjoying a life of "remarkably good health.")[12]

Although White, unlike other health reformers, grounded her sexual advice on revelation, not reason, in *Appeal to Mothers* she invoked both religious and scientific sanctions. She not only attributed her insights and advice to special revelation but sprinkled her text with Devil-talk, religious admonitions, and biblical citations. To Ellen White's thirty-four-page contribution, her publisher (probably her husband, James) appended a twenty-nine-page essay on "Chastity," which cited persons "of high standing and authority in the medical world" who agreed with her: Graham, Gove, Woodward, Jackson, the former Millerite physician Larkin B. Coles, and the phrenologist O. S. Fowler. With the exception of Woodward, this group of health reformers and hydropaths hardly represented the views of the medical elite, but given her low opinion of most physicians—"If there was in the land one physician in the place of thousands, a vast amount of premature mortality would be prevented," she had written— they were the most authoritative voices she could offer. So closely did the views of these individuals parallel those of Ellen White, the publisher of her booklet felt compelled to add a note denying that she had read their works before writing out what she had seen in vision. Taking her word at face value, he asserted: "[S]he had read nothing from the authors here quoted, and had read no other works on this subject, previous to putting into our hands what she has written. She is not, therefore, a copyist, although she has stated important truths to which men who are entitled to our highest confidence, have borne testimony."[13]

On October 2, 1868, over five years after her first view of the world's corrupt state, Ellen White received a second major vision on sex, which left her confidence in humanity "terribly shaken." As the sordid lives of "God's professed people" passed before her, she became "sick and disgusted with the rotten-heartedness" of her church. Protected by the cloak of divine revelation, she voyeuristically reported seeing reputable Adventist brethren leaving the "most solemn, impressive discourses upon the judgment" and returning to their rooms to engage "in their favorite, bewitching, sin, polluting their own bodies." Adventist children, she learned, were "as corrupt as hell itself." Speaking to the Battle Creek Seventh-day Adventist church in March 1869, she insisted, "Right here in this church, corruption is teeming on every hand." Privately, she estimated "that there is not one girl out of one hundred who is pure minded, and there is not one boy out of one hundred whose morals are untainted." Given the odds of any petitioner being a health-sapping masturbator, she decided to refuse future requests for prayers of healing.[14]

In addition to the multitudes who were abusing themselves, there were many others, she soon learned, who were abusing their spouses. In *How to Live* (1865), a set of bound pamphlets reporting various health-related aspects of her 1863 vision, she had urged couples to "consider carefully the result of every privilege the marriage relation grants," but until her 1868 vision she had focused on self-abusers, not spouse abusers. After 1868, however, she warned that even married persons were accountable to God "for the expenditure of vital energy, which weakens their hold on life and enervates the entire system." In phrenological language she counseled Christian wives not to "gratify the animal propensities" of their husbands but to seek instead to divert their minds "from the gratification of lustful passions to high and spiritual themes by dwelling upon interesting spiritual subjects." Husbands who desired "excessive" sex she regarded as "worse than brutes" and "demons in human form." In 1870 her husband brought out an expanded version of *Appeal to Mothers*, covering not only self- but spousal-abuse and published under the revealing title *A Solemn Appeal Relative to Solitary Vice, and the Abuses and Excesses of the Marriage Relation.*[15]

Ellen White's writings on sex received little publicity, in part because James, who for years presided over the church and controlled its publishing association, did little to promote them. Although James did write an extensive and laudatory introduction to *An Appeal to Mothers,* when it appeared, he simply listed it, without fanfare, among the publications available from the offices of the official church paper, at fifteen cents a copy.[16] In the early 1870s the *Health Reformer,* an Adventist magazine aimed at the general public, advertised books by Graham, Trall, and Jackson—and even White's *How to Live*—but not *Solemn Appeal.* James had long felt ambivalent about his wife's claims to divine authority for her views—for years in the 1850s he had denied her access to the official church magazine, *Review and Herald,* which he edited—and when he brought out *Solemn Appeal* in 1870, he stripped away all references to the visionary origins of his wife's views. According to one historian with unique access to Ellen White's private papers, "agonizing conflicts," sometimes resulting in physical separation, characterized the Whites' marriage from the mid-1860s until James's death in 1881.[17] One can only imagine how he felt about Ellen's coolness toward sex and her heartfelt condemnations of marital "excess."

Following the public recounting of the sex-oriented visions—which she called "testimonies"—in 1869 and 1870, some of which she published with the guilty identified by name, Ellen White wrote little on the subject for the rest of her life. She remained generally antipathetic toward sex, though she always stopped short of advocating celibacy. As far as I can determine, she never wrote a positive word about sex. In her waning years she looked forward expectantly to an idyllic existence in the New Earth, free from such unpleasant activities. When some members inquired in 1904 if there would be any children born in the next life, she replied sharply that Satan had inspired the question. It was he, she said, who was leading "the imagination of Jehovah's watchmen to dwell upon the possibilities of association, in the world to come, with women whom they love, and of their raising families." As for herself, she needed no such prospects. She died in 1915, a plump old woman at age eighty-seven, but her sexual advice has continued to circulate in various forms into the twenty-first century.[18]

John Harvey Kellogg

One of the primary reasons why White ceased to write about sex was that her protégé, John Harvey Kellogg, assumed her mantle. Young Kellogg began life in Michigan in a family newly converted to Seventh-day Adventism. His health-reforming parents subscribed to the *Water-Cure Journal* and tried to keep their sickly, puny son John out of the hands of physicians. At about age twelve John attracted the attention of James White, who lured him out of his father's broom factory to work in the Adventist press in Battle Creek. John stayed there for four years, rising from office boy to typesetter and proofreader. He set type for Ellen White's *How to Live* and, for a time, traveled and lived with the Whites as a virtual foster son. In his leisure he read the health-reform literature sold by the press, including, no doubt, Ellen White's recently published *Appeal to Mothers*. With the Whites' encouragement and financial assistance, he enrolled at age twenty in the Hygieo-Therapeutic College in New Jersey, a hydropathic doctor mill run by Russell Trall. Later he studied medicine at the University of Michigan Medical School and at the Bellevue Hospital Medical School, from which he received an M.D. in 1875. The next year, at age twenty-four, he became physician-in-chief at the failing decade-old Western Health Reform Institute, a Battle Creek water cure facility founded in response to one of Ellen White's visions. Kellogg promptly rechristened the institute the Battle Creek Sanitarium, coining the term "sanitarium" (as an alternative to the familiar "sanatorium") to designate "a place where people learn to stay well." The five-foot, four-inch dynamo, who for years dressed in white from head to toe, remained with the "San" for sixty-seven years, transforming it into a world-famous mecca for health seekers from around the globe.[19]

Kellogg's professional life, like his dual training in hydropathic and allopathic medicine, spanned the spectrum of medical respectability. He won fame not only for his array of hydrological, mechanical, and electrical treatments but also for his surgical prowess, especially in the areas of gynecological and gastrointestinal disorders, which Ellen White attributed to his always being assisted by an angel. He consorted with some of the leading physicians and surgeons of the Western world, as well as with such health faddists as Horace Fletcher, the great masticator. In 1886 Kellogg, an incorrigible publicity hound, barely escaped censure from his own county medical society for violating its ban against advertising and other infractions; yet two years later the group elected him president. He wrote (or dictated) nearly fifty books on topics ranging from the benefits of hydrotherapy and a meatless diet to the evils of masturbation and autointoxication (prevented by proper nutrition and frequent enemas). To keep the colon free of toxic bacteria, he recommended as many as five bowel movements a day.[20]

Despite his advocacy of temperance in all things, Kellogg found it impossible to live up to his own advice. Until late in life, he often slept only four hours a night and frequently skipped meals. At other times he gorged himself on vegetarian fare. He excused himself for such inconsistencies by noting his special role in life. "I am under no obligation to practice what I preach," he wrote privately. "My business is to preach and really I haven't time to practice. I am putting out fires; that is my business. I belong to the fire department, and I haven't time to look after my own health. I am looking after other people's health and my own health has to take the best chance it can." As an adult he suffered from chronic hemorrhoids and ulcers. In the twentieth century, when he broke down physically, reluctant to admit his own frailty, he would slip down to Florida to recuperate under the guise of taking a vacation or conducting business. To what extent he followed his own sexual advice can only be surmised. He repeatedly condemned masturbation as one of the greatest evils of the day, and he believed that ideally marital intercourse should be reserved for procreative purposes. It was widely rumored, however, that shortly before marrying Ella Eaton, a schoolteacher, he performed a hysterectomy on her, thus precluding any chance of progeny—and any opportunity to engage in sex for procreation. Perhaps, as John Money has suggested, Kellogg's frequent enemas substituted for sexual intercourse, as happens with people who suffer from a rare psychological disorder known as klismaphilia. Though biologically childless, the Kelloggs raised forty-two adopted or foster children.[21]

In the mid-1890s, working with his brother Will, John Kellogg serendipitously discovered flaked cereals—not to prevent masturbation, as is sometimes alleged, but because one of his patients had broken her false teeth on an earlier nutlike concoction. He also invented peanut butter, soy milk, and various coffee and meat substitutes. Although he lived well, he remained at heart a philanthropist, not a capitalist. He generously offered the right to his flaked cereals first to the Seventh-day Adventist Church and then, because Ellen White spurned them, to his brother Will. Despite his dietary lapses, Kellogg lived until 1943, when he died at age ninety-one, having influenced the eating habits and evacuation practices of millions. The *Journal of the American Medical Associa-*

tion, noting that Kellogg "held rigidly to a number of concepts, some of which did not meet with general medical approval," nevertheless hailed him as a "health evangelist."[22]

More than anything else, Kellogg produced books, magazines, and articles. He seldom went anywhere, even to the bathroom, without a stenographer or two in tow. Dictating at a rate of 180 to 200 words a minute, for hours at a time, he wrote books in very short order, reportedly relying only on brief outlines or his exceptional memory. His 133-page *Uses of Water in Health and Disease* (1876) took eight days; the 356-page *Plain Facts about Sexual Life* (1877), fourteen days. Never one to waste time, or to indulge in romantic frivolity, he added 156 pages to *Plain Facts* while on his honeymoon. According to his biographer, Richard W. Schwarz, the doctor once confided to his wife "that he did not believe it was necessary to get new material or even a new method of treatment or style when producing a new book; it was only necessary to give the appearance of a new book through changing the wording, chapter order, and headings." In this way Kellogg churned out over fifty books and more than a hundred pamphlets, one of which had a press run of a million copies. In addition, he wrote countless articles, edited several magazines, and delivered thousands of lectures. Few, if any, of his colleagues in the medical profession reached an audience as large as his, nationally or internationally.[23]

The publishing history of his magnum opus on sex typified his approach to bookmaking. First published in 1877 under the title *Plain Facts about Sexual Life*, when the sexually inexperienced author was only twenty-five years old, this 356-page book sold for $1.50 in cloth and 75 cents in flexible covers. By 1879 Kellogg had expanded it to 512 pages, by adding child-friendly chapters for boys and girls, and retitled it *Plain Facts for Old and Young*. During the 1880s under its new title it grew to 644 pages and in the 1890s to 720 pages. Finally, in 1917 Kellogg brought out a four-volume edition of over 900 pages, with extensive coverage of eugenics and race betterment, activities he had long promoted. A "wall of prejudice" against books on sex and the author's own obscurity initially kept sales depressed, but they picked up dramatically between 1879 and 1885, when nearly one hundred thousand copies were sold. After first publishing the book in Battle Creek, Kellogg turned to I. F. Segner, a former traveling bookseller, who had recently opened a bookstore and publishing company in Burlington, Iowa. We may never know why Kellogg turned to Segner, but it may be that explicit discussions of sex were a little too racy for the Adventists of Battle Creek, especially if not divinely revealed. By the opening of

the new century, sales had topped two hundred fifty thousand, and they eventually reached a reported five hundred thousand, a remarkable total (if accurate) for such a work.[24]

Although bestseller lists did not appear until 1895 and comparative figures
are virtually nonexistent, especially for nonfiction works, nineteenth-century
American books, even novels, were "considered highly successful," claims one
historian of publishing, "if 25,000 copies were sold, and a sale of 50,000 was as-
tonishing." William A. Alcott's very successful *Physiology of Marriage* (1856), for
example, sold over twenty-five thousand copies, as did Augustus K. Gardner's
Conjugal Sins against the Laws of Life and Health (1870). Frederick Hollick
boasted some three hundred "editions" of his *Male Generative Organs in Health
and Disease* (1849) and five hundred of his *Marriage Guide* (1850).[25]

The success of *Plain Facts* derived in part from squads of canvassing agents,
purportedly selected for their "discretion," who went door to door calling on
potential buyers in their homes. In recruiting agents for two other major books
by Kellogg, *Man, the Masterpiece* and *Ladies' Guide*, the publisher described the
recent evolution of colporteuring from "a sort of peddling business" into something that had "almost reached the dignity of a profession." Canvassers served
the public, by introducing them to valuable literature, and themselves, by providing a good income. *"There is no more pleasant, no more useful, and no more
profitable business* in which a young man or woman of ability can engage, *and
none which brings such large and quick returns* to the energetic worker, as the
sale of a good book in good territory," gushed the publisher. "A wide-awake
agent, with plenty of pluck and perseverance," could clear $25 to $50 a week,
earning a $2.75 commission per book. Such a person was not only an enter-
prising entrepreneur but "as genuine a missionary as the man or woman who
engages in missionary work in the wilds of Africa or the distant islands of
the seas."[26]

Kellogg, who had installed laboratories at the Battle Creek Sanitarium,
viewed himself as something of a scientist. In introducing himself to readers of
Plain Facts in the early 1880s, he noted on the title page that he held member-
ships in the American Public Health Association, the American Society [Asso-
ciation] for the Advancement of Science, the American Society of Microscopy
[Microscopists], and the Michigan State Board of Health, though his familiar-
ity stopped short of knowing their correct names. To ward off possible criticism, he warned readers in the preface that his "prime object" was "to call attention to the great prevalence of sexual excesses of all kinds, and the heinous

crimes resulting from some forms of sexual transgression, and to point out the terrible results which inevitably follow the violation of sexual law." But he quickly went on to say that in discussing such delicate subjects, he would use "the language of science [which] is always chaste in itself."[27] Throughout the text he invoked the names of various scientific and medical authorities to support his contentions. Although he displayed no reluctance to allude to religious sanctions and eternal salvation, he made no mention of his Adventist ties (beyond his connection to the Battle Creek Sanitarium) or his indebtedness to Mrs. White and her *Appeal to Mothers*. In content *Plain Facts* strayed little from the path marked out by White and the conventional views of contemporary advice writers.[28]

An 1882 printing of *Plain Facts* began with a low-key discussion of animal and vegetable sex, then moved on to sexual relations, chastity, continence, marital excesses, prevention of conception, infanticide and abortion, prostitution, and solitary vice, to which Kellogg devoted over a hundred pages of the most dire descriptions and warnings. Early on he established the intimate connection between nutrition and reproduction, "the two great functions of life," noting that the use of condiments in cooking promoted licentiousness and "unchastity." His reasoning went as follows: "The blood is made of what is eaten. Irritating food will produce irritating blood. Stimulating foods or drinks will surely produce a corresponding quality of blood. Irritating, stimulating blood will irritate and stimulate the nervous system, and especially the delicate nerves of the reproductive system." Graham or White could not have said it better. In addition to stimulating foods, he identified a host of other causes of unchastity: heredity, intermingling of sexes, overeating, tobacco, "bad books," idleness, fashionable dress, round dances, intestinal worms, overheated and underventilated rooms, and, especially among males, constipation, "one of the most general physical causes of sexual excitement."[29]

To maintain a chaste life, he recommended exercising the will and the body, adopting an abstemious diet, taking frequent baths, and embracing religion. Not surprisingly for a man who may never have engaged in sexual intercourse, he defended continence as a healthful and "natural" cure for unchastity. Like Ellen White, he excoriated husbands for destroying the health of their wives by making selfish sexual demands on them. He stopped short of calling for sexual abstinence when not intending to procreate, but he urged women to resist "the demands of the husband for the gratification of his bestial passions" and suggested that frequency of intercourse should be governed by the feelings of the

wife. He denounced contraception as a "crime against nature" that often resulted in the murder of a recently conceived human being. "It is evident," he
argued, "that at the very instant of conception the embryonic human being
possesses all the right to life it ever can possess. It is just as much an individual,
a distinct human being, possessed of soul and body, as it ever is, though in a
very immature form." For persons who found it difficult to limit themselves to
sex for reproduction only, he recommended separate beds or separate apart-
ments (an arrangement he and his wife used). He also offered a natural birth-
control "compromise" that had been circulating among health reformers for
years: restricting sexual activity to the period beginning fourteen days after the
onset of menses and continuing until three or four days before menstruation
begins again, during which conception was unlikely. He provided no supporting testimony of its efficacy.[30]

Kellogg devoted the most space and the gravest warnings to "the most dangerous of all sexual abuses," the practice of "solitary vice" (also known as "self-
pollution, self-abuse, masturbation, onanism, manustupration, voluntary pollution . . . secret vice, and other names sufficiently explanatory"). Quoting
Adam Clarke, a revered Methodist authority, Kellogg alleged that "neither the
plague, nor war, nor small-pox, nor similar diseases, have produced results so
disastrous to humanity as the pernicious habit of onanism." It destroyed not
only physical and mental health, but moral health as well, making salvation
impossible. Although Kellogg (like White) never addressed homosexuality directly, he did not, as Vern L. Bullough has suggested other nineteenth-century
writers did, conflate solitary vice and homosexuality. "As a sin against nature,"
he wrote in a rare reference to homosexuality, solitary vice "has no parallel except in sodomy (see Gen. 19:5, Judges 19:22)."[31]

In contrast to White, who identified the devil and his agents along with more
mundane factors as the prime culprits behind masturbation, Kellogg focused
exclusively on natural causes of the "disease" or "sin," terms he used inter-
changeably. In addition to inherited tendencies, tobacco, tea, coffee, "candies,
spices, cinnamon, cloves, peppermint, and all strong essences," he identified
"sexual precocity, idleness, pernicious literature, abnormal sexual passions, ex-
citing and irritating food, gluttony, sedentary employment, libidinous pictures,
and many abnormal conditions of life" as causes of the habit. His list of "sus-
picious signs" of masturbation ran even longer than White's: general debility;
early symptoms of consumption; premature and defective development; a sudden change in disposition; lassitude; an unnatural dullness and vacantness in

the eyes; sleeplessness; a failure of mental capacity; fickleness; untrustworthiness; love of solitude; bashfulness; unnatural boldness; mock piety; being easily frightened; confusion of ideas; aversion (or attraction) to the opposite sex; round shoulders; weak backs, pains in the limbs, and stiffness of the joints; paralysis of the lower extremities; a peculiar gait (moving stiffly for boys, wriggling for girls, shuffling for both sexes); bad positions in bed, such as lying on the abdomen or with hands on the genitals; a lack of development of the breasts in females; a capricious appetite (voracious in the beginning stages, impaired later on); extreme fondness for unnatural, hurtful, and irritating articles such as salt, pepper, spices, vinegar, mustard, and horseradish; the consumption of clay, slate pencils, plaster, and chalk; disgust for simple food; the use of tobacco; an unnatural paleness; acne, or pimples, on the face; biting the fingernails; a lack of luster and natural brilliancy in the eyes; a habitually moist, cold hand; palpitation of the heart; hysteria in females; chlorosis or green sickness; epileptic fits in children; wetting the bed; unchastity of speech and a fondness for obscene stories; regularly disappearing to secluded spots; and frequent seminal stains after puberty. Kellogg cautioned that only one or two of the above characteristics should not be regarded as diagnostic—that the only "absolutely positive signs" were frequent stains on nightshirts or sheets before a male child reached puberty, or an erect penis with a hand nearby, which could be discovered by slipping surreptitiously into the suspect's bedroom and quickly throwing back the covers.[32]

Kellogg never endorsed White's assertion that masturbation produced "cancerous humors," but he did believe that it could cause cancer of the womb, and, like her, he asserted that it caused a host of other medical problems: from urinary diseases, nocturnal emissions, and impotence to epilepsy and insanity, a commonly cited consequence of masturbation, especially since Samuel B. Woodward, superintendent of the Worcester Lunatic Asylum, began emphasizing the connection in the 1830s. Although females practiced self-abuse less frequently than males, they were more susceptible to masturbatory insanity than males and uniquely vulnerable to uterine diseases, atrophy of the breasts, and hysteria. As far as I know, Kellogg was among the first to note the possible psychological role of masturbation in producing insanity. Earlier in the century, asylum superintendents, especially in the Northeast, had ranked religious excitement among the leading causes of insanity. Although so-called religious insanity was disappearing as a diagnostic category during the last quarter of the century, Kellogg speculated that the psychological reaction to masturbation produced most cases of this type of mental derangement. "The individual is conscience-smitten in view of his horrid sins, and a view of his terrible condition—ruined for both worlds, he fears—goads him to despair, and his weakened intellect fails," he wrote; "reason is dethroned, and he becomes a hopeless lunatic."[33]

But *Plain Facts* carried a message of hope as well as one of fear. God could forgive sin, and Dr. Kellogg could cure disease. To combat the menace of secret vice, Kellogg prescribed an array of moral, behavioral, nutritional, medical, and surgical remedies. Parents should impress on their children the sinfulness of the habit and portray "in vivid colors its terrible results." They should make sure that their offspring were "fully occupied by work, study, or pleasant recreation." Young and old alike should adopt an abstemious diet, eating but twice a day and eschewing all stimulating foods and drinks. Though generally an advocate of strict vegetarianism, Kellogg allowed that small amounts of lean beef or mutton might be consumed in the early stages of recovery. As the author of a massive volume on hydrotherapy, he recommended a variety of water treatments: baths, douches, and fomentations. And as the apostle of colon hygiene, he stressed the importance of keeping the bowels clean, preferably without artificial assistance but with the help of enemas if necessary. "Useful as is the syringe when needed," he warned, "nothing could be much worse than becoming dependent upon it." Young children might be broken of the habit by bandaging their genitals, tying their hands, or covering their organs with a cage. Small boys almost always benefited from circumcision, especially as the surgeon operated without administering anesthesia. As Kellogg noted, "[T]he brief pain attending the operation will have a salutary effect upon the mind, especially if it be connected with the idea of punishment . . . The soreness which continues for several weeks interrupts the practice, and if had not previously become too firmly fixed, it may be forgotten and not resumed." Electrical shocks also proved helpful, "when skillfully applied." Under no circumstances, declared Kellogg, should drugs of any kind or mechanical devices such as rings or pessaries be used. "Quacks and charlatans," who found masturbators easy prey and who exploited their misfortunes for financial gain, should be shunned.[34]

How should we evaluate Kellogg's achievement as a purveyor of sexual advice? How can we even begin to assess the influence he had on sexual practices and attitudes in America—and indeed, around the world? Unlike his influence on the breakfast customs of Americans, which can be estimated by boxes of cereal sold, or his influence on the frequency of bowel movements, which might be reflected in the quantity of bran consumed or number of enema bottles manufactured, his influence on bedroom practices remains a slippery topic for historians. Birthrates in America did decline during the publishing lifetime of *Plain Facts,* but probably more because of the growing availability of contraceptives, which Kellogg condemned, than because of continence, which he advocated. Masturbation may have disappeared as a disease in the early twentieth century, but no evidence suggests that frequency of masturbation itself declined, Kellogg's sensationalist warnings notwithstanding.

Despite its immense circulation, *Plain Facts* failed to attract reviews in the leading journals of the day, and few references to it can be found in primary sources. I have found no record of readers claiming that their lives were improved—or ruined—by taking the doctor's advice, though Kellogg once spoke of the satisfaction he derived from "knowing in numerous instances that the virtue and happiness of whole families have been secured by the timely warnings of danger which parents have obtained from this work." He also told of hearing from scores of readers who expressed disappointment at not having had such guidance in their youths. "I would give all I possess in this world could I have had a copy of 'Plain Facts' placed in my hands when I was a lad," wrote one. "Words cannot express the gratitude I would now feel had some kind friend imported to me the invaluable information which this book contains; it would have saved me a life of wretchedness," lamented another. One anguished mother, whose institutionalized son had lost his mind through self-abuse, reportedly told an agent, "Oh, if I had only seen this work ten years ago my poor boy might have been saved!"[35]

Although Kellogg's writings on sex presumably influenced what many Americans believed about sex—and induced considerable guilt and fear in the minds of parents and children—the popularity of *Plain Facts* probably owed little to its actual impact on sexual practices. Aggressive marketing by a network of canvassers helped, as did Kellogg's unadorned, accessible style of writing. Unlike his mentor, Ellen White, whose strong claims to divine authority carried little weight beyond the small flock of Seventh-day Adventists, Kellogg, as a physician, could speak the universal language of medical science. Perhaps more important, as a Christian physician who shed his sectarian identity (and who was formally "disfellowshipped" from the Adventist church in 1907 for questioning White's prophetic gift), he could reassure his readers of the convergence of science and scripture on matters of sexual behavior. In contrast to

White, who had little of practical value to offer sexual sufferers beyond exhortations to exercise the will and adopt a bland vegetarian diet, Kellogg was able to add colon cleansing, his most distinctive practice, to the arsenal of preventive and curative measures against unwelcome sexual urges.

Both White and Kellogg may have scared readers out of their wits with their litanies of the dismaying moral and physical consequences of sexual activity. But to the extent that they offered the prospect of religious and medical salvation to those who would live by the laws of God and of nature, theirs was an optimistic message. Such persons might have to struggle against the oppression of sinful natures and hereditary predispositions, but their sins could be forgiven and their sicknesses cured.

NOTES

I am grateful to Jessica Kim, Libbie Freed, Richard Davidson, and Micaela Sullivan-Fowler for their assistance in the preparation of this essay.

1. James C. Whorton, *Inner Hygiene: Constipation and the Pursuit of Health in Modern Society* (New York: Oxford University Press, 2000), 182; William R. Hunt, *Body Love: The Amazing Career of Benarr Macfadden* (Bowling Green, Ohio: Bowling Green State University Popular Press, 1989), 45; Peter Gardella, *Innocent Ecstasy: How Christianity Gave America an Ethic of Sexual Pleasure* (New York: Oxford University Press, 1985), 45; John Money, *The Destroying Angel: Sex, Fitness, and Food in the Legacy of Degeneracy Theory, Graham Crackers, Kellogg's Corn Flakes, and American Health History* (Buffalo, N.Y.: Prometheus Books, 1985), 100 (masturbation); idem, "The Genealogical Decent of Sexual Psychoneuroendocrinology from Sex and Health Theory: The Eighteenth to the Twentieth Centuries," *Psychoneuroendocrinology* 8 (1983): 398 (nonsense). See also M. E. Melody and Linda M. Peterson, *Teaching America about Sex: Marriage Guides and Sex Manuals from the Late Victorians to Dr. Ruth* (New York: New York University Press, 1999), 27–35.

2. Hugh M. Hefner, "The Playboy Philosophy: Part 17," *Playboy* 11 (June 1964): 29–39, 111–15, quotation on 29; T. Coraghessan Boyle, *The Road to Wellville* (New York: Viking, 1993). Regarding the movie, see Ronald L. Numbers, review of "The Road to Wellville," *Journal of the History of Medicine and Allied Sciences* 50 (1995): 283–85.

3. See, for instance, James C. Whorton, *Crusaders for Fitness: The History of American Health Reformers* (Princeton, N.J.: Princeton University Press, 1982), esp. 201–38.

4. See, for instance, Richard W. Schwarz, "John Harvey Kellogg: American Health Reformer" (Ph.D. diss., University of Michigan, 1964), published with abridgments and without documentation as *John Harvey Kellogg, M.D.* (Nashville, Tenn.: Southern Publishing Association, 1970); and Gerald Carson, *Cornflake Crusade* (New York: Rinehart, 1957). In *John Harvey Kellogg, M.D.*, Schwarz devotes only a few sentences (88, 150) to Kellogg's *Plain Facts* and ignores its content; Carson neglects even to mention the book.

5. Stephen Nissenbaum, *Sex, Diet, and Debility in Jacksonian America: Sylvester Graham and Health Reform* (Westport, Conn.: Greenwood Press, 1980); Sylvester Graham, *A Lecture to Young Men on Chastity . . .* (Providence, R.I.: Weeden and Cory, 1834).

6. William A. Alcott, *The Young Man's Guide* (Boston: Samuel Colman, 1835); [Mary Gove], *Solitary Vice: An Address to Parents and Those Who Have the Care of Children* (Portland, Maine: The Journal Office, 1839); [Samuel B. Woodward], *Hints for the Young, on a Subject Relating to the Health of Body and Mind* (Boston: Weeks, Jordan, 1838); Samuel Gregory, *Facts and Important Information for Young Women, on the Subject of Masturbation; with Its Causes, Prevention, and Cure* (Boston: Geo. Gregory, 1845); William A. Alcott, *The Physiology of Marriage* (Boston: Dinsmoor, 1855); Russell T. Trall, *Pathology of the Sexual Organs, Embracing All Forms of Sexual Disorders* (Boston: B. Leverett Emerson, 1862); James C. Jackson, *The Sexual Organism, and Its Healthful Management* (Boston: B. Leverett Emerson, 1861), 11.

7. Ronald L. Numbers, *Prophetess of Health: A Study of Ellen G. White* (New York: Harper & Row, 1976), reprinted in a revised and enlarged edition as *Prophetess of Health: Ellen G. White and the Origin of Seventh-day Adventist Health Reform* (Knoxville: University of Tennessee Press, 1992), on which this section heavily rests. All citations are to the latter edition. For additional biographical information, see also Jonathan M. Butler, "Prophecy, Gender, and Culture: Ellen Gould Harmon [White] and the Roots of Seventh-day Adventism," *Religion and American Culture* 1 (1991): 3–29; and Ronald D. Graybill, "The Power of Prophecy: Ellen G. White and the Women Religious Founders of the Nineteenth Century" (Ph.D. diss., Johns Hopkins University, 1983).

8. Numbers, *Prophetess of Health*, 77–85, 150–53. Regarding Edson, see Graybill, "Power of Prophecy," 62–68.

9. Ellen G. White, *An Appeal to Mothers: The Great Cause of the Physical, Mental, and Moral Ruin of Many of the Children of Our Time* (Battle Creek, Mich.: Seventh-day Adventist Publishing Association, 1864), 17–18, 24–25, 27–28. In appearance and tone, White's *Appeal to Mothers* looked much like hydropath Mary Gove Nichols's eighteen-page pamphlet *Solitary Vice*.

10. White, *Appeal to Mothers*, 5–10, 14, 19–20.

11. Ibid., 11–12. On White's own feelings about sex, see Numbers, *Prophetess of Health*, 221.

12. Ellen G. White, *Special Testimony for the Battle Creek Church* (Battle Creek, Mich.: Seventh-day Adventist Publishing Association, 1869), 21; "C. L. Daigneau, Old Resident of City, Dies," *Benton Harbor, Mich., News-Palladium*, 13 February 1928, 8; "Body of Late S. E. Daigneau Is Here for Burial," ibid., 8 July 1931, 3. I am indebted to Jean Davis of Battle Creek for information about the Daigneau family, who lived a couple of houses away from the Whites and whose son Samuel was about the same age as Edson White. When Ellen White first published what she had seen in vision about the Daigneau family, she mentioned the parents and children by name; later, when this testimony appeared in her collected *Testimonies for the Church*, 9 vols. (Mountain View, Calif.: Pacific Press, n.d., [ca. 1868]), 2:404, the family members were identified solely by sequential letters of the alphabet.

13. Ellen G. White, *Spiritual Gifts: Important Facts of Faith, Laws of Health, and Testimonies Nos. 1–10* (Battle Creek, Mich.: Seventh-day Adventist Publishing Association, 1864), 133 (physicians); White, *Appeal to Mothers*, 34.

14. White, *Testimonies for the Church*, 2:349–50, 360, 439, 468–69; 4:96.

15. Ellen G. White, *Health; or, How to Live* (Battle Creek, Mich.: Seventh-day Adventist Publishing Association, 1865), 2:48; White, *Testimonies for the Church,* 2:472–75; James White, ed., *A Solemn Appeal Relative to Solitary Vice, and the Abuses and Excesses of the Marriage Relation* (Battle Creek, Mich.: Seventh-day Adventist Publishing Association, 1870).

16. *Advent Review and Sabbath Herald* 24 (1864): 176.

17. Numbers, *Prophetess of Health,* 28; Graybill, "Power of Prophecy," 25.

18. E. G. White, letter B-59-1904, quoted in J. E. Fulton, *Pacific Union Recorder* 31 (July 1932): 2, quoted in Numbers, *Prophetess of Health,* 159. The Seventh-day Adventist Church still officially defends White's teachings about sex; see, for example, the appendix on "Masturbation and Insanity" in a recent anthology, *Testimonies on Sexual Behavior, Adultery, and Divorce: A Compilation from the Writings of Ellen G. White* (Silver Spring, Md.: Ellen G. White Estate, 1989), 268–70; and Herbert E. Douglass, *Messenger of the Lord: The Prophetic Ministry of Ellen G. White* (Nampa, Idaho: Pacific Press, 1998), 493–94. On the development of a less negative attitude toward sex among some Adventists, beginning in the early 1930s, see Malcolm Bull and Keith Lockhart, *Seeking a Sanctuary: Seventh-day Adventism and the American Dream* (San Francisco: Harper & Row, 1989), 133–35; and Ronald L. Numbers and David R. Larson, "The Adventist Tradition," in *Caring and Curing: Health and Medicine in the Western Religious Traditions,* ed. Ronald L. Numbers and Darrel W. Amundsen (New York: Macmillan, 1986), 447–67, esp. 458.

19. Schwarz, *John Harvey Kellogg, M.D.,* 2–13, 17, 25, 27–32, 59, quotation on 62.

20. Ibid., 35–36, 46–58, 87–94, 109–15. On the history of autointoxication, see Micaela Sullivan-Fowler, "Doubtful Theories, Drastic Therapies: Autointoxication and Faddism in the Late Nineteenth and Early Twentieth Centuries," *Journal of the History of Medicine and Allied Sciences* 50 (1995): 364–90.

21. Schwarz, *John Harvey Kellogg, M.D.,* 128–35, 151; Money, *Destroying Angel,* 84. According to Kellogg's biographer, Richard W. Schwarz (personal communication, 2 August 1998), it was widely gossiped in Battle Creek Adventist circles that Kellogg had performed a hysterectomy on his fiancée shortly before marrying her. His source was Mary Lamson, a private teacher the Kelloggs hired for their family.

22. Schwarz, *John Harvey Kellogg, M.D.,* 116–20, 209–11; Numbers, *Prophetess of Health,* 188–89; obituary in the *Journal of the American Medical Association* 123 (1943): 1132.

23. Schwarz, *John Harvey Kellogg, M.D.,* 88–89; Robert G. Cooper, "A Comprehensive Bibliography of Dr. John Harvey Kellogg, 1852–1943" (Keene, Texas: mimeographed, 1984).

24. Major editions include *Plain Facts about Sexual Life* (Battle Creek, Mich.: Office of the Health Reformer, 1877); *Plain Facts about Sexual Life* (Battle Creek, Mich.: Good Health Publishing, 1879); *Plain Facts for Old and Young* (Burlington, Iowa: Segner & Condit, 1879); *Plain Facts for Old and Young: Embracing the Natural History and Hygiene of Organic Life* (Burlington, Iowa: I. F. Segner, 1885); *Plain Facts for Old and Young: Embracing the Natural History and Hygiene of Organic Life* (Burlington, Iowa: Segner, 1895); *Plain Facts for Old and Young; or, The Science of Human Life from Infancy to Old Age: A Cyclopedia of Special Knowledge for All Classes on the Hygiene of Sex,* 20th century ed. (Battle Creek, Mich.: Health Library Association, 1901); *Plain Facts for Old and Young; or, The Science of Human Life from Infancy to Old Age* (Battle Creek, Mich.: Good Health Publishing, 1903); *Plain Facts,* 4 vols. (Battle Creek, Mich.: Good Health Publishing,

1917). An advertisement in John Harvey Kellogg, *The Household Manual: Of Hygiene, Food and Diet, Common Diseases, Accidents and Emergencies, and Useful Hints and Recipes* (Battle Creek, Mich.: Office of the Health Reformer, 1877), 173, describes the first edition of *Plain Facts*. Kellogg refers to the "wall of prejudice" in the introduction to the 1885 edition. Schwarz, *John Harvey Kellogg, M.D.*, 88, reports total sales of half a million. Other sales figures are taken from prefaces and advertisements. *Plain Facts* is currently available online from the University of Virginia Library: http://etext.lib.virginia.edu. On Segner, see his obituary in the *Burlington, Iowa, Gazette*, 29 January 1917, 7, a copy of which I obtained through the courtesy of Susie Guest of the Burlington Public Library.

25. John Tebbel, *A History of Book Publishing in the United States*, 4 vols. (New York: R. R. Bowker, 1972–82), 2:68; [William A. Alcott], *The Physiology of Marriage* (Boston: J. P. Jewett, 1856); Augustus K. Gardner, *Conjugal Sins against the Laws of Life and Health* (New York: J. S. Redfield, 1870); Frederick Hollick, *The Male Generative Organs in Health and Disease, from Infancy to Old Age* (New York: Nafis & Cornish, 1850); idem, *The Marriage Guide; or, Natural History of Generation* (New York: T. W. Strong, 1850); Alice Payne Hackett, *70 Years of Best Sellers, 1895–1965* (New York: R. R. Bowker, 1967), 27. No books on medicine or sex appear on Hackett's list of "Early Best Sellers" (235–38), which comprises books published before 1895 that sold more than a million copies. Regarding I. F. Segner, see "Died in Kansas City: Mr. I. F. Segner Passed Away in Missouri City Today," *Burlington, Iowa, Gazette*, 29 January 1917, 7; copy courtesy of Susie Guest, Burlington Public Library. Some evidence suggests that Kellogg's success remained unequaled until 1932, when Max J. Exner published *The Sexual Side of Marriage*, which enjoyed sales of over a million during the next thirty-three years. The tendency to inflate publishing statistics is illustrated by the eclectic physician Edward Bliss Foote, who claimed on the title page of the 1887 edition of his *Plain Home Talk* (originally published in 1870) that 500,000 copies had been sold, while simultaneously telling a correspondent that the figure was 250,000; see Janet Farrell Brodie, *Contraception and Abortion in Nineteenth-Century America* (Ithaca, N.Y.: Cornell University Press, 1994), 202.

26. J. H. Kellogg, *Plain Facts for Old and Young* (Burlington, Iowa: I. F. Segner, 1882), vi; advertisement placed by Modern Medicine Publishing Company on the back cover of *Good Health* 28 (January 1893), recruiting agents to sell Kellogg's *Ladies' Guide in Health and Disease: Girlhood, Maidenhood, Wifehood, Motherhood* (Des Moines, Iowa.: W. D. Condit, 1883) and *Man, the Masterpiece; or, Plain Truths Plainly Told about Boyhood, Youth and Manhood* (Des Moines, Iowa: W. D. Condit, 1885). Boldface in original.

27. Schwarz, "John Harvey Kellogg," 149–51; Kellogg, *Plain Facts for Old and Young* (1882), v–vi.

28. In "American Attitudes on Sexual Hygiene and Ethics, 1877–1914: A Study of Ideology in the Works of John Harvey Kellogg, Periodical Literature, and Manuals of Advice" (master's thesis, California State University, San Francisco, 1973), George Corcoleotes found few differences between Kellogg and his fellow advice givers.

29. Kellogg, *Plain Facts for Old and Young* (1882), 30, 182–204, 210.

30. Ibid., 205, 209–15, 244, 247, 257, 260–61, 264–67.

31. Ibid., 315–17. In "Homosexuality and Its Confusion with the 'Secret sin' in Pre-Freudian America," *Journal of the History of Medicine and Allied Sciences* 28 (1973): 143–55, Vern L. Bullough and Martha Voght argue that Victorian writers commonly conflated homosexuality with masturbation.

32. Kellogg, *Plain Facts for Old and Young* (1882), 321–46.

33. Ibid., 371, and in general "Results of Secret Vice," 347–77. On religion and insanity, see, for example, Ronald L. Numbers and Janet S. Numbers, "Millerism and Madness: A Study of 'Religious Insanity' in Nineteenth-Century America," in *The Disappointed: Millerism and Millenarianism in the Nineteenth Century,* ed. Ronald L. Numbers and Jonathan M. Butler (Bloomington: Indiana University Press, 1987), 92–117. See also [Samuel B. Woodward], *Hints for the Young, on a Subject Relating to the Health of Body and Mind* (Boston: Weeks, Jordan, 1838).

34. Kellogg, *Plain Facts for Old and Young* (1882), 383–84, 406–7, 412. See also J. H. Kellogg, *Rational Hydrotherapy* (Philadelphia: Davis, 1901).

35. Kellogg, *Plain Facts,* 24; J. H. Kellogg, *The Home Hand-Book of Domestic Hygiene and Rational Medicine* (Battle Creek, Mich.: Good Health Publishing, 1886), 358.

CHARLES E. ROSENBERG (volume editor) is Professor of the History of Science and Ernest E. Monrad Professor of the Social Sciences at Harvard University. He has published widely in the social history of medicine and has had a long-time interest in popular medicine.

KATHLEEN BROWN is an associate professor of history at the University of Pennsylvania and the author of *Good Wives, Nasty Wenches, and Anxious Patriarchs: Gender, Race, and Power in Colonial Virginia* (1996). She is currently completing a study of early American cleanliness practices titled *Foul Bodies and Infected Worlds: Cleanliness in Early America*.

MARY E. FISSELL teaches in the Department of the History of Science, Medicine, and Technology at the Johns Hopkins University. She is working on a cultural history of *Aristotle's Masterpiece*.

WILLIAM H. HELFAND has written, lectured, and exhibited extensively on the history of drugs and pharmacy, and on prints, caricatures, posters, and ephemera relating to pharmacy and medicine.

THOMAS A. HORROCKS is Associate Director for Special Collections and the Joseph Garland Librarian at the Francis A. Countway Library of Medicine, Harvard University. He is pursuing a doctorate in history from the University of Pennsylvania. His dissertation, titled *Rules, Remedies, and Regimens: Health Advice in Early American Almanacs*, examines how print both influenced and was influenced by popular views of the body, health, and disease in pre–Civil War America.

RONALD L. NUMBERS is Hilldale and William Coleman Professor of the History of Science and Medicine and chair of the Department of Medical History and Bioethics at the University of Wisconsin, Madison. He is the author of

Prophetess of Health: A Study of Ellen G. White (1976) and coeditor of *Medicine without Doctors: Home Health Care in American History* (1977), *Sickness and Health in America: Readings in the History of Medicine and Public Health* (1978), and *Caring and Curing: Health and Medicine in the Western Religious Traditions* (1986).

STEVEN SHAPIN teaches in the Department of Sociology and Science Studies Program at the University of California, San Diego. His books include *A Social History of Truth* (1994) and *The Scientific Revolution* (1996).

JEAN SILVER-ISENSTADT holds a Ph.D. in the history of science from the University of Pennsylvania and an M.D. from the University of Maryland School of Medicine. She is author of *Shameless: The Visionary Life of Mary Gove Nichols* (2002).

STEVEN STOWE teaches history at Indiana University, Bloomington. He is, most recently, editor of *A Southern Practice: The Diary and Autobiography of Charles A. Hentz, M.D.* (2000) and is completing a study tentatively titled *Doctoring the South: Physicians and Illness in the Mid–Nineteenth Century American South.*